T0146022

On the Other Hand

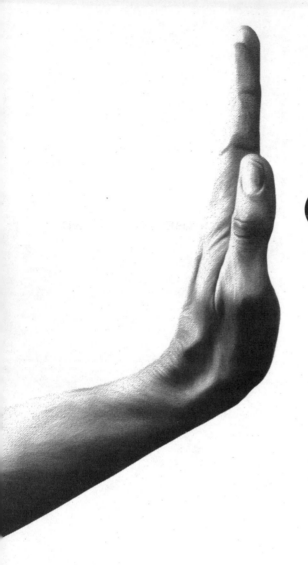

On the

Other Hand

Left Hand,
Right Brain,
Mental Disorder,
and History

Howard I. Kushner

Johns Hopkins University Press · Baltimore

© 2017 Johns Hopkins University Press
All rights reserved. Published 2017
Printed in the United States of America on acid-free paper
9 8 7 6 5 4 3 2 1

Johns Hopkins University Press
2715 North Charles Street
Baltimore, Maryland 21218-4363
www.press.jhu.edu

Library of Congress Cataloging-in-Publication Data
Names: Kushner, Howard I., author.
Title: On the other hand : left hand, right brain, mental disorder, and history /
 Howard I. Kushner.
Description: Baltimore, Maryland : Johns Hopkins University Press, 2017. |
 Includes bibliographical references and index.
Identifiers: LCCN 2016052579| ISBN 9781421423333 (hardcover : alk. paper) |
 ISBN 9781421423340
Subjects: | MESH: Functional Laterality | Mental Disorders—etiology | Attitude
Classification: LCC RC386.6.N48 | NLM WL 335 | DDC 616.8/0475—dc23
 LC record available at https://lccn.loc.gov/2016052579

A catalog record for this book is available from the British Library.

Special discounts are available for bulk purchases of this book. For more information,
please contact Special Sales at 410-516-6936 or specialsales@press.jhu.edu.

Johns Hopkins University Press uses environmentally friendly book materials,
including recycled text paper that is composed of at least 30 percent post-
consumer waste, whenever possible.

To the two left-handers who shaped my life
in nonsinister ways:

Gertrude Slotnikoff Klein (1919–2002)

Carol Rose Rubin Kushner

Contents

Preface

Strangely, there is little agreement as to precisely how handedness should be defined or measured, yet nearly all people readily identify themselves as being either right-handed or lefthanded.

MICHAEL CORBALLIS (1997)

Left-handers are a lot like Canadians. They immediately recognize one another. My interest in left-handedness, as psychiatrists used to say, is overdetermined: a combination of personal experience overlaid with clinical observations. It began with my own left-handedness. The lack of left-handed desks in my grammar school no doubt contributed to my inability to master the cursive writing skills expected of second-graders in the 1950s. As a result, my penmanship grades remained in the basement for my entire grammar school career. Nevertheless, my teacher, Miss Coyle, predicted that despite my handicap, I would become a writer. Otherwise my left-handedness presented no drawbacks in my childhood. I was an acceptable first-baseman and a useful Little League switch-hitter. Technology saved me. My early attempts at writing took place on an electric typewriter and advanced to a KayPro DOS computer, of which I was an early and eager user.

A generation earlier was quite different. A natural left-hander, my mother was forced to write and sew with her right hand, a practice enforced by educators in the Philadelphia public schools in the 1930s. Unlike some other left-handers who were forced to switch, my mother did not stutter, though she had a number of reading difficulties that she masked but could not overcome. And she was markedly unco-ordinated, which she attributed to her forced hand switching. Even today on most of planet Earth, left-handed children continue to be "retrained" as right-handers and as a result have reading and some-times speaking disabilities.[1]

The second impetus for this study was clinical, based on my obser-vations of pediatric patients at a movement disorders clinic at Memo-rial Hospital of Rhode Island, Brown University School of Medicine, in 1994 and 1995. For a book on Tourette syndrome I was working on,[2] the clinic's director, pediatrician Louise Kiessling, graciously invited me to observe patients and participate in a research project investigat-ing streptococcal antibodies as a possible cause of Tourette syndrome.

The left-handed author, age 3,
with his left-handed mother.
(Author's personal photo.)

During my time at Brown I noticed that the patients with Tourette syndrome and attention deficit disorder (ADD/ADHD) appeared to be left-handed in a greater proportion than the 10 percent of the population expected to be left-handed. To humor me, Kiessling and two other colleagues, Linda Abbot and Joe Hallett, agreed to use the standard Edinburgh Handedness Inventory to determine how many of our patients were indeed left-handed. My hypothesis was confirmed, but the findings were statistically inconclusive due to our small sample size and lack of controls.

I had planned to undertake a full-blown research project on handedness and pediatric movement disorders. The project was sidetracked, however, by my work with a team of colleagues at the University of California at San Diego on the etiology of Kawasaki disease, a sometimes-fatal pediatric vasculitis. In the fall of 2000 I moved to Emory University, where along with the Kawasaki project I worked on addictions and self-medication by people with chronic diseases, especially those diagnosed with pediatric movement disorders. Again, because a disproportionate number of this population appeared to be left-handed, I was propelled to consider the role of handedness in learning disorders. More than a decade and a half after my experiences at Brown, I resolved to stay focused on the mysteries of left-handedness.

But there were more than 15 years' worth of research and publications awaiting my reading. I assumed that many of the questions I had asked in the 1990s must have been answered while I was AWOL from the topic of left-handedness, so I was surprised to learn that the controversies seemed only to have multiplied, despite research performed by some extraordinarily talented and highly trained researchers. As I discovered, almost every question about the extent, origin, function, and consequence of left-handedness remained either contested or unresolved.[3]

Despite the vast number of articles published on the topic and wide public interest in it, only a few books published in the past half century have tried to provide a comprehensive exploration of the causes and consequences of left-handedness. Five of these are aimed at a popular audience. Michael Barsley's 1966 *The Other Hand: A Social History of Left-Handedness*, chock-full of entertaining vignettes, is aimed at exposing the historical discrimination against left-handers.[4] Along with his three other similar book-length examinations, in this book Barsley sought to dispel myths of left-handedness and to provide a positive sense of self-identity for left-handers. As is often the case, Barsley's corrective reinforced other myths, including the claim that left-handers were more creative and talented than right-handers.[5]

In contrast to Barsley's positive spin on left-handers, British Columbia psychologist Stanley Coren's 1992 book *The Left-Hander Syndrome: The Causes and Consequences of Left-Handedness* emphasized the downside of left-handedness. For Coren, preference for the right hand was normal, while left-handedness, in his words, was "right-handedness run amok." According to Coren, left-handedness resulted from "a failure to reach right-handedness."[6] Citing his published collaborative research, Coren attached left- and mixed-handedness to "abnormal development" of the left hemisphere, which in turn increased risk for "dyslexia, attention deficit disorders, learning disabilities, and mental retardation." Purposely provocative, Coren's *The Left-Hander Syndrome* gained a wide popular audience. Less enthusiastic, academic investigators were particularly skeptical about Coren's claim and data alleging that left-handers died younger than right-handers. Coren's interpretations, published more than a quarter of a century ago, have not always worn well in the context of the subsequent outpouring of research and publication on handedness. Nevertheless, Coren's book serves as an important example of an influential strand of late twentieth-century thinking about left-handedness.[7]

Rejecting both Barsley's positive take on left-handedness and Coren's pathologizing is British neurosurgeon and psychologist Chris McManus's *Right Hand, Left Hand: The Origins of Asymmetry in Brains, Bodies, Atoms, and Cultures.* A tour de force, McManus's book, drawn from his years of investigation into handedness, is comprehensive, authoritative, and persuasive. Using a multidisciplinary approach, McManus examines the biological and cultural underpinnings of asymmetry, destroying the many myths of its origins.[8] This nuanced account is required reading for serious researchers and general scholars.[9]

Drawing on a wide range of sources from the Middle Ages to the mid-twentieth century, art historian Pierre-Michel Bertrand's *Histoire des gauchers: Des gens à l'envers* (2001) provides a chronology of attitudes and practices toward left-handers moving from hostility to tolerance (or indifference) and, ultimately, to admiration and pride. Bertrand's history, however, is not one of unmitigated progress, but rather a narrative of struggle, resilience, and contradiction. A valuable resource, Bertrand's book is unavailable in English and in subsequent chapters I have included his ideas and insights in my discussions.[10]

As I was completing my book manuscript, a new book—*Laterality: Exploring the Enigma of Left-Handedness*—by the respected psychologist and researcher Clare Porac was released. A professor at Pennsylvania State University and former collaborator with Stanley Coren, Porac aims to translate the "vast and frequently contradictory" academic studies on left-handedness into "accessible and readable form." Concentrating on twenty-first-century studies, Porac relies mainly "on meta-analyses and review articles," especially those by "well-known and highly respected" researchers.[11] The book is valuable for academic specialists; given its topical format and focus, general readers will probably find it as useful as a reference book for particular topics of interest as a text to read from cover to cover.[12]

Current disagreements over the causes and consequences of left-handedness mirror, albeit with a different vocabulary, medical, anthropological, and scientific debates dating to the early twentieth century. Sophisticated tools and technologies have not reduced the influence of the stigmatization of left-handedness—stigmatization that has hounded left-handers since early human history. For this reason

I reexamine left-handedness in the context of the growing field of disability studies that has contested the classifications and meanings of disability, forcing researchers to reexamine their assumptions and attitudes about disability while challenging public policies aimed at them.[13]

One thesis of my book is that the history of left-handedness parallels that of disabilities. Thus, left-handedness was understood as a sign of abnormality and of the profane, which was opposed to the normal or the sacred. Attitudes and practices toward left-handers reflect both transcendent and culturally specific values.

A second thesis is that the damage produced by discrimination against left-handers is much greater than the putative pathology resulting from left-handedness. By the early twentieth century the rhetoric of traditional stigmatization of left-handers was replaced by a scientific, social, and educational discourse that (re)authorized forcing left-handers to become right-handed. Although forced hand switching is now rare in the West, it continues to be widely practiced in China, India, and much of Africa. Moreover, the assumption of so many investigations that left-handedness is connected to learning disabilities has contributed to the belief that left-handers are abnormal.

Examining left-handedness in historical perspective provides a crucial context for current controversies, not only about the causes and consequences of left-handedness, but also about the relationship, if any, between handedness, linguistic laterality, and learning disabilities. This history also reveals my third thesis: that toleration of left-handedness serves as a barometer of wider cultural toleration and permissiveness. Societies and cultures that discriminate against left-handers are also less tolerant of other forms of diversity.

ON THE OTHER HAND

1

GENES AND KANGAROOS

The robust, species-wide lateralization that exists in humans is unusual, and perhaps unique among primates. . . . [There are] several possible evolutionary explanations for this strong asymmetry. In particular . . . that preexisting hemispheric asymmetry for perception and language processing drove the evolution of human handedness.
FITCH AND BRACCINI (2013)

Bipedal macropod marsupials display left-forelimb preference at the population level in a variety of behaviors in the wild. In eastern gray and red kangaroos, we found consistent manual lateralization across multiple behaviors. This result challenges the notion that in mammals the emergence of strong "true" handedness is a unique feature of primate evolution . . . [and] leads to the conclusion that the interspecies differences in manual lateralization cannot be explained by phylogenetic relations, but rather are shaped by ecological adaptations.
GILJOV ET AL. (2015)

I undertook this history of left-handedness knowing that I would encounter many genetic and environmental hypotheses that purported to explain why most humans relied on their right hand for manual tasks and their left cerebral hemisphere for speech and language. This is referred to as asymmetry, literally the lack of equality (of symmetry), in terms of size, shape, and function. Laterality describes the left or right direction of the asymmetry. In the vast majority of humans, language and speech are lateralized to their left cerebral hemisphere, which is also referred to as their dominant hemisphere.

Asymmetries are found throughout the natural and animal world, from proteins (left) and sugars (right) to mice and human hearts, viscera, and livers. In most humans the heart is lateralized left and the liver is lateralized right. Despite this universal asymmetry, there is no agreement about whether the brains of human primates, let alone other species, are asymmetrical and lateralized to the same extent as humans.

I also anticipated examining the putative links between atypical asymmetry (left-handedness and right-brain dominance) and learning disabilities and talent. I never expected kangaroos. It turns out that bipedal species of kangaroos not only display a left-forelimb preference but also, lacking a corpus callosum—that bundle of neurons connecting the cerebral hemispheres—resemble some persons with autism, a disorder sometimes connected to both left-handedness and an underdeveloped corpus callosum. These findings about kangaroos have been used as evidence for both environmental claims and genetic claims about the etiology of left-handedness.

Researchers have drawn on the best science of their eras in seeking to explain what causes left-handedness, but as the example of kangaroo laterality reveals, the same evidence is open to different interpretations. The history of interpretations of both the causes and effects of left-handedness leads inevitably to the present, with its still tentative, still inexact understandings. The current search for a molecular explanation of handedness, to take one example, has a very long history, punctuated by periodic announcements that the genetics of handedness has been identified. Through all the different eras, experts

Red tree kangaroos, a bipedal species that displays a left-forelimb preference. They have been cited as evidence for both environmental claims and genetic claims about the etiology of left-handedness. (From http://www.zooborns.com /zooborns/2014/05/tree-roo-roger -williams.html.)

found these and announcements of other explanatory models to be premature or incomplete.

Since the early twentieth century most investigators of hand-edness, influenced by their disciplinary training, have sought their answers in either culture or biology but rarely in both. This binary approach, although cloaked in the rhetoric of science, also reflects an older belief that right- and left-handedness represented a con-flict between the sacred and profane. This dualism is evident in early twentieth-century evolutionary medicine, where left-handedness was attributed to putatively less "evolved" populations, especially non-Europeans. Its diagnostic signs included criminality, disease, and mental retardation.[1] Despite the evidence that men were more likely to be left-handed than women, left-handedness was gendered "female" and attached to women and other so-called primitives.[2] Right-handedness, in contrast, was viewed as a sign of advanced civi-lization and health.[3] There has been surprisingly little attention paid to handedness among African Americans. These studies, which we will examine in detail later, have reported a higher prevalence of left-handedness in African Americans compared to European Americans.[4] Additionally, left-handedness has been gendered "gay," while ambi-dexterity has been linked to lesbianism.[5]

In contrast to those working in evolutionary medicine, a num-ber of French sociologists and anthropologists, along with an eclectic group of British social reformers, advocated the view that left-handers were a suppressed minority, which if freed from discrimination would display great talent and creativity.[6] Indeed, the British ambidexter-ity movement of the early twentieth century insisted that ending discrimination against left-handers would liberate right-handers as well, because they too would have access to both hands and both ce-rebral hemispheres. Freed from old prejudices and constraints, these new ambidextrals would become intellectual and artistic leaders. The controversy between those who view left-handedness as a health risk and those who view it as a sign of creativity and talent persists today among researchers and the public. Although a few observers have adopted a dialectic approach in which health risks as well as talent potential can be found among left-handers,[7] the old dualisms remain surprisingly resilient.

A historical exploration of understandings about handedness forces us not only to seek new research tools but also to confront the need to develop more robust classification systems. Continuing attempts to identify the gene or genes implicated in handedness illustrate this process.

CAUSES: MEDIA, MOLECULES, AND MESSAGE

In this age when the cause of behaviors and traits increasingly is assumed to be genetic, investigations into left-handedness have focused on searching for the putative gene or genes that determine handedness. Although genetic hypotheses have been proposed since the early twentieth century, today's genomics relies on sophisticated statistical and imaging tools that take advantage of the recent elaboration of the human genome.

These studies often gain wide, if fleeting, media coverage. Science writer Rik Smits points to a now forgotten 2007 BBC report proclaiming that a British team had identified "the gene for left-handedness," which also increased the risk for mental illnesses, including schizophrenia.[8] As is often true when the media translate scientific hypotheses, the complexities and caveats of researchers are frequently sacrificed to the demands of provocative headlines and accessible summaries. The studies themselves are not bad science. Indeed, like the study I discuss below, they are methodologically sophisticated and intriguing. But identifying the etiology of handedness always turns out to be much more complex than it appears, not least of all because there is, as the respected New Zealand psychologist Michael Corballis writes, no agreement about how to define left-handedness.[9]

Investigators, including myself, are frequently attracted to left-handedness because it has been connected to learning disabilities such as mental retardation, autism, dyslexia, schizophrenia, Tourette syndrome, attention disorders, and stuttering.[10] The classification and definition of these disabilities, however, is fluid and highly contested, so reliable data are difficult to develop and initial findings often impossible to replicate.[11] Given these constraints, only the most courageous researchers set out to uncover the causes of handedness and then to connect them with the origins of learning and language disabilities.

Among them is a team of British and Dutch researchers who, according to a wide variety of media reports, in September 2013 had discovered the genes for left- and right-handedness. Typical was the story in *Medical News Today* proclaiming that "scientists have uncovered genes that are involved in determining whether a person is left or right handed."[12] As is generally the case, the study itself, published in *PLOS Genetics,* was more tentative. Its title, "Common Variants in Left/Right Asymmetry Genes and Pathways Are Associated with Relative Hand Skill," reveals a more modest finding than that reported in the media. An examination of this study suggests that caution was certainly in order.

The team led by Oxford University researcher William M. Brandler hypothesized that the genes responsible for handedness were the same ones that underlay the early development of asymmetry and laterality in the rest of the body.[13] Employing a method called Genome-Wide Association Study (GWAS), Brandler and colleagues searched for gene variants that might be connected with "hand skill." Their GWAS identified a candidate gene, *PCSK6,* an enzyme that turns on a protein identified with embryonic growth in mice and other mammals. The molecular mechanism of *PCSK6* had been characterized as playing a crucial role for establishing left and right asymmetry in a wide array of species. Mice bred to eliminate *PCSK6* displayed a number of asymmetry disorders, including abnormal position of their organs. Irregularities of the *PCSK6* allele, it was hypothesized, interfered with normal asymmetric development in humans.[14] The Brandler study concluded that disruption of *PCSK6* should be more evident in left-handers than in right-handers.

To test their hypothesis the team examined 728 people with dyslexia for hand skills, assuming that those with this reading disability would be more likely to be left-handers than the normal population.[15] They employed a hand-skill test to identify left-handed people who have dyslexia. In the GWAS of their dyslexic population Brandler and colleagues uncovered a number of other candidate genes that might be associated with disruptions in left/right asymmetries but were convinced that disruption of *PCSK6* was strongest. Given these mixed results, the team concluded that handedness was determined by a number of polymorphisms, including *PCSK6.* Moreover, they conceded that their investigation had not identified "the precise relationship between handedness, cerebral asymmetry, and neurodevelopmental disorders like dyslexia."[16] "As with all aspects of human behaviour," they wrote, "nature and nurture go hand-in-hand." Had the media reported the article accurately, their headline should have been, "British/Dutch Team Reports Left-Handedness Results from a Combination of Genes, Environment, and Culture." But then, there was no news in such a headline.

Brandler proposed a larger follow-up focused on the genetic substrates of dyslexia and other language-related disorders to overcome the limitations of the study and tease out the developmental connections, if any, between handedness and language.[17] In the subsequent investigation Brandler and his collaborator, Silvia Paracchini, examined dyslexics for disruptions of the *PCSK6* gene. Their study was based on two controversial assumptions. The first, building on Brandler's earlier study, was that handedness was determined by the same biological mechanisms that established other left/right asymmetries. The second was that handedness and brain asymmetry were connected to dyslexia and schizophrenia.

As to the first assumption, Brandler and Paracchini noted that *PCSK6* had been connected to the left/right direction of cilia, the hairlike projections of some cells. Cilia are lateralized left. Since cilia are found in organisms from snails to humans, some researchers have hypothesized that the genes that code for cilia also code for more complex asymmetries in humans such as organ placement and handedness.[18]

The second assumption, that dyslexia and schizophrenia were more common in left-handers, required an even greater leap: that the mutant form of the gene for cilia not only coded for left-handedness but also was responsible for dyslexia and schizophrenia. Brandler and Paracchini concluded that "the genetic nature of this correlation is not understood."[19] Given this limitation, theirs is a bold hypothesis based, as it is, on a number of controversial correlations, not least of all the definition and classification of dyslexia, schizophrenia, and left-handedness.

Other geneticists were skeptical of the Brandler claims. Writing in *Heredity*, University of Nottingham geneticist John Armour and his colleagues rejected Brandler's link between *PCSK6* alleles and handedness with learning disabilities because the Brandler study had not been replicated or even tested on a normal control population. Although Armour and his colleagues also assumed that handedness was hereditary, they nevertheless insisted that the nature and mechanisms of the inheritance remained elusive. Many genetic models "have been proposed that fit different features of the observed resemblance between relatives," Armour insisted, "but none has been decisively tested or a corresponding causative locus identified." Pointing to their analysis of 3,940 twins, Armour and his colleagues reported that they had been unable to discover any gene locus associated with handedness. Thus they concluded that their data failed to validate the type of relationship suggested in the Brandler studies.[20]

In a subsequent interview, Armour was less restrained in his criticisms of the Brandler hypothesis. Although there were numerous "weak genetic factors in handedness," Armour maintained that there were no strong ones. It would require much larger populations than even those of his twin investigations, Armour said, to identify these putative genes. But, he concluded, "even if these genes are identified in the future, it is very unlikely that handedness could be usefully predicted by analysis of human DNA."[21] And recently a Dutch study reported that a linkage analysis of the wide array of heritable traits revealed no major gene for either left-handedness or for atypical language lateralization.[22]

Brandler and Paracchini's assumption of a connection between dyslexia and left-handedness was equally questionable. According to the widely respected and authoritative Oxford University psychologist Dorothy V. Bishop, there is no evidence to support the assumption that dyslexia and similar disorders are connected with left-handedness. Citing her extensive literature search, Bishop "found no association between handedness and language and literacy impairments."[23] Thus we should adopt the same skepticism toward the interpretation of a widely reported recent study claiming left-forelimb preference among a slight majority of tree kangaroos described at the beginning of this chapter. The authors have cited the finding as evidence for the role of evolutionary pressures of bipedalism in etiology of hand preference because no similar preference was found among quadruped marsupials.[24] "The neural basis of manual lateralization found in bipedal marsupials," wrote the authors, "is unknown." But, they assert, it "is an important subject for future research."[25] They expect that such research will provide clues for human laterality that others have tied to bipedalism.[26]

Because marsupials lack a corpus callosum, some in the popular media have once again linked left-handedness to autism. Their reasoning is based on the claim that the corpus callosum is underdeveloped (hypoplastic) or absent (agenesis) among persons with autism. In fact, some reports have even suggested that bipedal kangaroos display autistic signs.[27] Thus left-handedness among humans may serve as a proxy for both an underdeveloped corpus callosum and autism. However, a recent meta-analysis* of all the studies cited on the PubMed website from 2000 to 2013 connecting autism and left-handedness does not support such a conclusion, not least of all because of the studies' lack of reliability. That is, although these studies examine autism or left-handedness, for instance, they define them differently. Thus combining the data of several studies is often problematic.[28]

Like so many other earlier announcements of the discovery of the cause of handedness, the Brandler studies and the interpretation of the putative kangaroo forelimb preference ultimately lack reliability and validity and, for the reasons I will examine below, these limitations cannot be easily repaired by more studies based on similar assumptions. Nevertheless, these studies are the latest entries in a long and ongoing debate over whether handedness is inherited or learned and whether left-handedness is associated with learning disabilities.

EFFECTS: SICKER OR SMARTER?

Brandler's use of dyslexia as a substitute, or proxy, for left-handedness provides a window into approaches to thinking about the effects of left-handedness, the second issue that has consumed both popular and scientific studies. Not "What causes left handedness?" but rather "Are left-handers at greater risk of disease than right-handers or are they exceptionally talented and creative?"

Searching "left-handedness" online leads to seemingly contradictory claims. The first attaches left-handedness to learning disabilities, especially, but not limited to, autism, ADHD, Tourette syndrome, stuttering, dyslexia, mental retardation, and schizophrenia. For instance, approximately simultaneously with the publication of the Brandler

---

* A meta-analysis is a retrospective examination of independent studies that in combination provide a large sample population that can be subjected to statistical analysis. This is a common approach in handedness studies, which we will examine throughout this book.

studies, in mid-November 2013, a *Time* magazine blog proclaimed, "Handedness may be a window into biological markers that predict psychoses." The blogger referenced a recent study at the Yale Child Study Center[29] as suggesting that left-handers "are more highly represented among those suffering from psychotic disorders such as schizophrenia."[30]

Alternatively, there is a widespread belief that left-handers are smarter, more talented, or more creative than right-handers. The site Anything Left-Handed posits that there are "more left-handed people with IQs over 140 than right-handed people—which is the 'genius' bracket." This, according to the site, explained why there were more left-handers "in creative professions—such as music, art and writing—and more left-handed astronauts and leaders than would be expected."[31] A 2006 study by the National Bureau of Economic Research in Cambridge, Massachusetts, claimed that left-handed men earn 5 percent more than their right-handed counterparts.[32] The study, which has never been replicated, was front-page news two years later in the *London Daily Mail*.[33]

A popular slogan among advocates of left-handed superiority, available on mail-order T-shirts, is "Only left-handed people are in their right brains."[34] Despite the pun on the word "right," the slogan exhibits a number of misconceptions about handedness and about the brain. Because *motor* functions are contralateral—that is, the left hemisphere motor cortex controls the right hand for right-handers, while for left-handers it is the right hemispheric cortex that controls the left hand—the assumption has long been that for right-handers language is located in the left hemisphere, while for left-handers language resides in the right hemisphere. But it is not true.

Over the years popular media and numerous books, including the bestseller *Drawing on the Right Side of the Brain*, which has gone through four editions between 1979 and 2012, have bolstered the perception that the right hemisphere is the seat of creativity and talent.[35] In combination these popular sources have given rise to the slogan and pun that left-handers are in their "right" brains. Left-handedness has become a convenient substitute for investigations of language and cognition in the right hemisphere. But, again, it is not true.

Contrary to popular perceptions, left-handers are not more linguistically right-brained than right-handers. An array of imaging technologies, including functional magnetic resonance imaging (fMRI) and functional transcranial doppler imaging (fTCD) have allowed re-searchers to examine hemispheric language placement in large healthy populations. These investigations have revealed that left-handers are *not* normally right-brained for language and cognition. In fact, only 18 percent of left-handers are right-brained for language, and 5 per-cent of right-handers are right-brained for language. Given that there are at least 10 to 12 right-handers for every left-hander, right-brained right-handers vastly outnumber right-brained left-handers. "If left-handers can be left-brain dominant, and right-handers can be right-brain dominant," asks author Melissa Roth, "then what do our hand skills really tell us about our brains?"[36]

Relying on handedness as a proxy for linguistic dominance, as most studies for the past century have done, misleads us about the nature and etiology of human handedness as well as the possible relationships between hemispheric dominance and language disor-ders. As neuroscientists W. Tecumseh Fitch and Stephanie Braccini have recently argued, "The link between handedness and linguistic lateralization is tenuous at best."[37] Similarly, German neuroscien-tist Sebastian Ocklenburg and his colleagues question the long-held assumption that handedness and hemispheric laterality share a common etiology. They insist that "the strength of this correlation depends on the measures used to assess the two traits, and the neuro-physiological basis of language lateralization is different from that of handedness." Although these two traits share a partially common genetic substrate, they arise from independent forces.[38] Endorsing this conclusion, Bishop has called for the elimination of handedness as a proxy for cerebral lateralization.[39]

Moreover, the presumption that the two hemispheres have com-pletely separate tasks is widely held but ultimately misleading, if not erroneous. The influential British neurosurgeon and psychologist Chris McManus writes that the assignment of different tasks to each hemisphere, the left for language and the right for nonverbal tasks, is "pleasingly straightforward"—and wrong. It is a big mistake to assume "that everyone's brain is the same," says McManus, who rejects these characterizations of the brain and language as "simplistic."[40] As he reminds us, presumed right-hemisphere processes are not always in the right hemisphere and left-hemisphere processes are not in the left hemisphere. Some linguistic mechanisms, according to McManus, such as speech, reading, writing, and spelling, are not always found together in the same hemisphere.[41] This is especially important in any discussion of what is referred to as language dominance.

Handedness, Language, and Other Asymmetries

Of course, motor functions generally are contralateral. Thus, there are a number of disorders, such as stuttering, dyslexia, and Tourette syndrome, which, though they affect speech or reading, may be primarily related to motor signaling and timing. In these disorders handedness *may* serve as a reasonable proxy for motor laterality.[42]

Humans are asymmetrical for a variety of functions, including seeing, hearing, and foot use, but for only a few tasks do they display species-wide dominance, as in hand use, where the asymmetry favors the right hand rather than the left. Other motor functions are asymmetrical but not robustly lateralized to the left or right hemisphere. For example, hearing and seeing, though asymmetrical in humans, are not nearly as lateralized to one side as is hand use in eating, writing, or throwing, which are strongly influenced by cultural values. Hearing laterality is just about equally split between the right and left hemispheres, unlike handedness, in which 90 percent of right-handers are left-brained for eating and throwing. Indeed, increasing evidence suggests that human hearing and seeing preferences developed independently and for different reasons from hand preferences. Seeing and hearing seem to operate more efficiently in the species if asymmetrical, but it might not matter whether the dominance is located on one side or the other.

Although language, speech, and handedness are located in the left hemisphere for the vast majority of right-handers, left-handers are right-hemisphere dominant for motor functions but not for language and speech. Thus, for the vast majority of left-handers, like right-handers, language and speech are located in their left hemisphere. This suggests to Cambridge University neuroscientist R. E. Rosch and colleagues that the etiology of handedness is different from that of speech and language. Moreover, although language and handedness are asymmetrical and lateralized in the same hemisphere—again, right-handers draw on the left hemisphere for language, and left-handers the reverse—it does not necessarily follow that they are dependent or even connected with one another. Thus Rosch challenges the long-held assumption of "a single underlying factor that leads to left hemisphere language and right hemisphere visuospatial processing in the majority of people." Based on fTCD imaging, Rosch found no correlation in individuals between visual and language laterality in either "direction" or "in laterality index size" (810). This study and others like it reinforce the increasing evidence that there are "multiple independent" mechanisms that determine linguistic and cognitive lateralization. These studies call into question long-held assumptions about the evolution of human laterality.[43] Fitch and Braccini conclude that there is "little evidence of causal links between handedness and other forms of laterality."[44]

Even here, the extent of the dominance (laterality) of right-handed asymmetry depends upon which set of tasks one examines. Among literate populations writing is most often the task examined, but one can easily recognize the limitations of using this behavior, because it has been so widely attached to social taboos and cultural pressures. The same may be said for the eating hand in less literate societies. Indeed, when one examines the literature carefully there is great diversity in determining handedness, especially in those societies where writing and eating are intimately connected to cultural values so that left-handers are often forced to switch to their right hands for culturally laden tasks.[45] As many critics have noted, the instruments used to measure handedness, particularly the widely used Oldfield/Edinburgh and the Annett Handedness inventories, assume that all the left-right asymmetries found in humans should be given equal weight in determining the extent of right- or left-handedness, but as we have seen this is a questionable assumption.[46]

LINGUISTIC LATERALITY AND HANDEDNESS

Disentangling linguistic laterality from human handedness forces us to reexamine and possibly resolve the contradictions and anomalies found in the literature of brain asymmetry. Brandler and colleagues believe that there is a common gene that codes mammalian embryos, including humans, for left-right asymmetries. Thus they insist that handedness and language asymmetry—in fact all mammalian asymmetries—have a common genetic origin.[47] But most other experts are skeptical of this claim.

Corballis adds a layer of complexity, hypothesizing that human hemispheric language specialization may have its origins prior to primate evolution and thus prior to the development of language itself. If language asymmetries preceded human handedness laterality, reasons Corballis, language hemispheric specialization most likely developed from human communication rather than from manual dexterity.[48]

Venturing beyond Corballis, Bishop suggests a neuroplasticity model that turns on its head the standard assumption that handedness determines language laterality. She hypothesizes that the "causal path between cerebral lateralization and language and literacy development is reversed; that is, language ability influences cerebral lateralization." Warning that the current focus on identifying laterality genes has resulted in ignoring nongenetic factors, Bishop offers two alternative potential causal mechanisms. The first is a "systematic influence" such as "the position of the fetus in utero" or a number of small environmental and parental influences that in combination result in both motor and linguistic asymmetries. Such perturbations could and probably do influence the asymmetries of higher cognitive functions. Alternatively, she suggests that the entire process might result from random gene expression.[49]

Supporting Bishop's hypotheses are the data, discussed earlier, that reveal that only 18 percent of left-handers are linguistically right-brained. In fact, while fewer than 6 percent of all humans are linguistically right-brained, the majority of the linguistically right-brained are right-handers. These data provide additional evidence against the validity of widespread use of handedness as a proxy for linguistic laterality. In fact, most infants begin to speak at 9 to 12 months but do not settle on a dominant hand until almost their third year or later. This opens the possibility that linguistic sidedness might determine handedness. Indeed, a number of recent genetic models of handedness—such as those of Marian Annett and Chris McManus— assume that hand preferences arise during early gestation and are linked to linguistic sidedness. Only later on, approximately at age 3, is handedness determined by skill—or, alternatively, cultural and environmental pressures.[50] If so, this raises a difficult problem for the identification of handedness itself. Should handedness be a measure of preference or of skill?[51] If it is preference, then the overwhelming majority of investigations are based on invalid data. That is because

handedness studies almost always focus on individuals and populations older than 3 years of age. As a result, they measure a combination of preference and skill. Because genetic studies like Brandler's
are based on reported handedness, they cannot lead us to putative
genetic substrates, because we have no agreed method to measure
hand preference in populations.[52] When we combine this fact with
the limitations we have already noted in handedness surveys, we see
that if we can't agree on what we should measure then it is impossible
to determine its extent, let alone its cause.

Classifying Learning Disabilities

The claims that the same mechanisms that result in left-handedness
underlie learning disabilities are also problematic. Learning disabilities are syndromes rather than diseases. In contrast, measles, polio,
smallpox, and sickle cell disease are diseases because a tentative diagnosis based on signs and symptoms is confirmed or rejected through
a laboratory test indicating infection by a pathogen or the presence of
a genetic mutation. The cause of a syndrome remains unknown.[53] Because neither the etiology nor pathophysiology of syndromes has been
identified, a diagnosis of autism, Tourette syndrome, schizophrenia, or
ADHD depends on identifying a list of possible combinations of signs
and symptoms displayed by a patient within a specified time frame
in contrast to a control population. The list of signs and symptoms
is tentative, and disagreement often surfaces over which signs and
symptoms are crucial to authorize a diagnosis.[54] Recognizing the tentative nature of a syndrome can be productive, because it authorizes
researchers and clinicians to question underlying assumptions about
the signs and symptoms and to provide alternative hypotheses for the
etiology of idiopathic disorders. Uncritical and exclusive adherence to
a list of signs and symptoms can pose a risk, because atypical cases
often provide insight into possible etiological features. What is seen
as a single clinical entity may result from a variety of different underlying mechanisms, or one underlying mechanism may manifest itself
in a variety of signs and symptoms.

The definitions and diagnoses of learning disabilities have wid-
ened over time, making it difficult to determine whether the incidence and prevalence have changed or whether earlier cases were

not diagnosed.[55] This problem is especially evident in the controversies surrounding the changes in the 5th edition of the *Diagnostic
and Statistical Manual of Mental Disorders* (*DSM-5*) regarding autism,
specifically over whether the classification of pervasive developmental
disorders (PDD)—which includes a spectrum from severe impairment to high-functioning persons now diagnosed as having Asperger's
syndrome—should be retained.[56] Similarly, the diagnostic criteria for
Tourette syndrome have been contested for more than 40 years.[57] The
uncertainties about what constitutes these syndromes pose a great
challenge for those who wish to examine the relationship between
learning disabilities and left-handedness. Unfortunately, many researchers have simply ignored these classification issues, often making it difficult to compare their findings with the results emerging
from other studies of handedness and specific disorders.[58]

Despite the initial promise of a breakthrough, contemporary attempts
to identify the genetic origins of left-handedness continue to disap-
point. In addition, a number of extremely creative genetic models
remain controversial. Although few researchers today would support
the proposition that handedness is solely the result of education, most
would probably agree that environmental factors play a role in determining left-handedness. As Bishop recently suggested, there may
be a variety of routes, rather than one common etiology, to human
handedness.[59] A similar conclusion was reached by the respected
Scottish educator Margaret M. Clark more than half a century ago in
her classic study *Left-Handedness: Laterality Characteristics and Their
Educational Implications.*[60] Such a multifactorial model might help
untangle the contradictory observations and data supposing that left-
handers are at greater risk for learning disabilities and the assertions
that left-handers are more talented and creative than right-handers—
suppositions that lead us inevitably to the topic of the stigmatization
of left-handers.

    This tension between characterizing left-handers as abnormal and
disabled versus insisting that they are talented victims of discrimination has deep historical roots. These conflicting positions are best
illustrated by the early twentieth-century debate between the Italian
criminologist Cesare Lombroso and the French anthropologist Robert
Hertz, the subject of the next chapter.

2

CRIMINALS OR VICTIMS?

Cesare Lombroso vs. Robert Hertz

Ghosh meant to ask Bachelli if he actually believed anything in Lombroso's abominable book, La Donna Delinquente. *Lombroso's "studies" of prostitutes and criminal women uncovered "characteristics of degeneration"—such things as "primitive" pubic hair distribution, an "atavistic" facial appearance, and an excess of moles. It was pseudoscience, utter rubbish.*

ABRAHAM VERGHESE (2009)

Hertz's thesis was remarkable for its time, and even today its value is obvious. . . . Constant attempts have been made to exploit it and to extend and improve its use. Criticism has generally been muted and indirect, to such an extent that one feels that Hertz himself has taken on a rather sacred quality for many.

ROBERT PARKIN (1996)

I n spite of my mother's learning deficits, when I was growing up I was convinced that left-handers were more talented, if a bit weirder, than right-handers. I also assumed, with no empirical data, that Jews were more likely to be left-handed than gentiles. My heroes were left-handed first-basemen, but none that I knew of were particularly smarter than their right-handed teammates—nor were any Jewish. Still, professional baseball was one place where left-handers were admired and, though probably more common in baseball than in other professions, they were, like me, a minority. In any case, I remained convinced that left-handers were exceptionally talented and often Jewish. This may have been reflected in my choice of friends; certainly, I immediately noticed other left-handers.

Given my assumptions I connected prejudice against left-handers with anti-Semitism. I had no idea that my beliefs were addressed in the early twentieth-century controversy between the Italian criminologist Cesare Lombroso (1835–1909) and the French anthropologist

Robert Hertz (1881–1915) over whether left-handedness was the cause of disability or the seat of talent; over whether left-handers were criminals or victims of discrimination. Lombroso had connected left-handedness with a variety of pathologies, including criminal behaviors and mental retardation, while for Hertz lifting the stigma attached to left-handedness would unleash long-suppressed creative potentials. In both cases, but for different reasons, their interpretation of left-handedness was influenced by their fears of anti-Semitism.

Although almost every overview of left-handedness cites the views of Lombroso and Hertz, few have examined the motivations for their claims and the contexts in which they made them. Like the hypotheses that would be offered by experts in the century that followed, the nineteenth-century competing views of Lombroso and Hertz reflected differences in disciplinary assumptions about what constitutes valid evidence—in this case, to attribute the causes and consequences of

Cesare Lombroso (1835–1909), the Italian criminologist who connected left-handedness with a variety of pathologies, including criminal behaviors and mental retardation. (Photo courtesy of the Wellcome Library via Wikimedia Commons.)

Robert Hertz (1881–1915), the French anthropologist who advocated that eliminating the prejudices toward left-handers would unleash the dammed-up energy of the right hemisphere of the brain, increasing human talent and creativity. (Photo courtesy of Bibliothèque de sciences humaines et sociales Descartes CNRS.)

left-handedness to biology or to culture. Additionally, Lombroso's and Hertz's claims served as proxies for deeply held assumptions about what constitutes race and whether or not populations are capable of behavioral change. Their own Jewish ancestry and European anti-Semitism played unacknowledged but powerful roles in both men's formulations of both "race" and left-handedness. Although Lombroso's criminal anthropology indicated that behavior was inherited and generally immutable, Lombroso nevertheless contended that any criminal activities among Jews was a response to persecution, not biology.[1] In contrast Hertz drew a parallel between the historical prejudice toward and discrimination against left-handers and Jews. Conceding that handedness had a biological origin, Hertz, in a 1909 essay, "The Pre-Eminence of the Right-Hand: A Study on Religious Polarity," nevertheless insisted that biology was trivial compared to the devastating impact of discrimination against left-handers. Ending stigmatization of left-handers would unleash the dammed-up energy of the right hemisphere of the brain, increasing human talent and creativity.[2] Ending discrimination of Jews would have a similar impact. Although later researchers would not connect stigmatization of left-handers with anti-Semitism, their explanations often reflected other subterranean prejudices and assumptions that framed the science of the time. Because their formulations continue to influence present attitudes toward left-handedness, it is productive to examine how Lombroso and Hertz arrived at their explanations.

THE BORN CRIMINAL AND THE LEFT SIDE

According to Lombroso, criminals and mental defectives could be identified by their physical features and behaviors, which resembled those of nonhuman primates and "savages."[3] He used the word "atavism" to label this regression to primitivity. Lombroso's claims were elaborated in five editions of *Criminal Man* (1876 to 1897)[4] and in his companion book, *Criminal Woman, the Prostitute, and the Normal Woman* (1893), written with his disciple Guglielmo Ferrerro.[5]* In his

---

* Lombroso's views on the causes of criminality were complex and included the recognition of environmental forces along with and sometimes separate from heritable factors.

publications Lombroso aimed to create a scientific criminology based on the quantification and classification of atavistic traits, key among them left-handedness.

Lombroso elaborated the connections between criminality, insanity, and mental retardation and left-handedness in his 1903 article "Left-Handedness and Left-Sidedness."[6] Although left-handers did not necessarily evince all the anatomical characteristics found among criminals, left-handedness served as evidence for Lombroso of the physical manifestation of their atavism. As "man advances in civilization and culture," wrote Lombroso, "he shows an always greater right-sidedness as compared to savages, the masculine in this way outnumbering the feminine and adults outnumbering children [who] even when they are not properly left-handed have certain gestures and movements which are a species of left-handedness." That, explained Lombroso, "was why in early times, and still among people little civilized, such as Arabs, the writing was preferably from right to left, which is the habit of children until corrected."[7]

Left-handers, according to Lombroso, have a greater sensitivity to the left side and the right cerebral hemisphere. Combining left-handers with left-siders, Lombroso formulated the hypothesis that left-handers were more likely to be criminals and that lunatics were generally left-sided (441, 443). Although Lombroso did not claim that all left-handers were criminals or insane, he was convinced "that left-handedness, united to many other traits, may contribute to form one of the worst characters among the human species" (444).

Though not without critics, Lombroso's formulation was influenced by and resonated with contemporary legal and scientific thinkers.[8] Lombroso drew on widely articulated claims that criminality and insanity were connected with weakened laterality.[9] Among "primitives," according to Lombroso, the right hemisphere (thus, the left side) was favored over the left hemisphere (the right side). He agreed with the French anthropologist Gustave LeBon (1841–1931), who claimed that non-right-handedness was a sign of failed asymmetry and thus of mental inferiority.[10] LeBon located "failed asymmetry" more common in "savages" and women, as evident from their smaller skull size: "Whilst the average size of the skulls of male Parisians places them among the largest known skulls, the average size of those of female Parisians places them among the very small skulls observed, very much below those of Chinese women and scarcely above those of the women of New Caledonia."[11] Similarly, according to Lombroso's colleague psychiatrist Enrico Morselli (1852–1929), insanity and criminality resulted when the primitive, emotional, and feminine right hemisphere overwhelmed the rational left hemisphere.[12]

Lombroso also found confirmation of his views in "degeneration theory." Expounded by the French physician Bénédict Augustin Morel (1809–1873) and psychologist and philosopher Théodule Ribot (1839–1916), degeneration theory offered a hereditarian explanation for a variety of disorders including retardation, depression, depravity, and sterility.[13] Behaviors, including addictions to alcohol and criminality, were alleged to have a cumulative destructive impact on the nervous system that was inherited by succeeding generations.[14] Degeneration theory was the soil in which Lombroso's atavism took root. Degeneration was, Daniel Pick notes, "a shifting term produced, inflicted, refined, and reconstituted in the movement between human sciences, fictional narratives and socio-political commentaries."[15] If for some of its French adherents identification of degeneration provided the first and necessary step for its amelioration, for Lombroso it confirmed the futility of attempts to cure or even mitigate degeneracy. That was because for Lombroso criminals were natural primitives, not so much unwilling as unable to adopt modern civilized rules of behavior. It was "pointless to pontificate on the moral responsibility of atavistic individuals, but crucial to separate them out from the rest of society," he wrote.[16]

This conclusion presented a dilemma for Lombroso, since it seemed to provide ammunition for the claims of late nineteenth-century anti-Semites. Lombroso, who was descended from a prominent Jewish family, was extremely sensitive to the arguments that Jews were a separate race and could never fully assimilate into European society. Thus, when it came to analyzing criminal behavior among Jews, Lombroso jettisoned his atavistic and biological arguments in favor of a cultural explanation. Lombroso insisted that Jewish behavior was a response to persecution rather than from biological inheritance. Lombroso asserted that Jews who turned to criminal behaviors did so in response to "desperate poverty" and their historical "need for protection against persecution." If Jews often served "feudal lords as receivers of stolen goods," it was, wrote Lombroso "to avoid being massacred." Given these social and political pressures Lombroso was surprised that Jewish criminals were not more numerous. Differentiating them from left-handers, Lombroso noted that Jewish criminal behaviors decreased wherever political participation was open to them.[17]

ROBERT HERTZ AND THE CULTURE OF HANDEDNESS

In 1909, the year of Lombroso's death, a young French sociologist, Robert Hertz, challenged Lombroso's claims in a *Revue Philosophique* article titled "The Pre-Eminence of the Right Hand." "Lombroso believes himself to have justified scientifically the old prejudice against left-handed people," wrote Hertz, who insisted instead that the predominance of the right-handed was deeply ingrained in prehistoric beliefs about the sacred and the profane.[18]

The profane was not merely a negative characteristic, it also was an "antagonistic element that by its very contact degrades, diminishes, and changes the essence of things that are sacred."[19] This dualism, according to Hertz, is evident in every aspect of social organization. It reinforces social hierarchies in which the lower social orders are portrayed as unclean and restricted from contact with the noble classes, who are assigned sacred duties (95). These distinctions are represented in the human body itself, in which the male is designated as the right side while the female is associated with the left side (98). Thus the male/right becomes synonymous with power and vigor, in contrast with the female/left, which is associated with weakness and passivity. Although Hertz acknowledged that these associations most likely reflected typical brain asymmetry in which the left hemisphere dominates, making the right hand generally preferred, he nevertheless rejected the claim that predominance of the right hand was biological. Rather, Hertz argued, left-brain dominance resulted from the ancient cultural assignment of superiority to the right side and right hand. "In spite of appearances," wrote Hertz, "the testimony of nature is no more clear or decisive, when it is a question of ascribing attributes to the two hands, than in the conflict of races or the sexes" (89). Thus, Hertz claimed that "if organic asymmetry had not existed, it would have had to be invented" (98).

Hertz not only aimed to reveal the origins of prejudice against left-handers, he also sought to liberate left-handers from the ancient discriminatory practices that accompanied them. The restrictions on the use of the left hand, he noted, also were aimed at repressing individuality: "The systematic paralysation of the left arm has, like other mutilations, expressed the will animating man to make the sacred predominate over the profane, to sacrifice the desires and the interest of the individual over the demands felt by the collective consciousness" (112). Hertz believed these ancient prejudices should not restrain modern humans from expressing their individuality, particularly their choice of handedness.

Hertz supported his views through an examination of the rituals, symbols, and behaviors of aboriginal peoples. He drew on a number of anthropological studies but mainly those of ethnographer Elsdon Best (1856–1931), who studied the Tuhoe tribe of the Maori people of New Zealand.[20] Lombroso had asserted that in "primitive" cultures, populations tended to be predominantly left-handed. Hertz set out to demonstrate that human societies, contrary to Lombroso's claims, had always been predominantly right-handed.

This cultural and linguistic dualism was obvious among the Maori and other aboriginal peoples.[21] In all these societies the right hand is associated with the holy, and the left with illegitimacy and the occult. "A left hand that is too gifted and agile is a sign of a nature contrary to the right order, of a perverse or devilish disposition: every left-handed person is a possible sorcerer, justly to be distrusted" (106). These assumptions about the nature of the left-handed explained to Hertz why there are so few of them: society favors right-handers, and education is aimed at "paralyzing the left hand while developing the right" (106). As Hertz described earlier in his essay, to ensure they would not use their left hands, native children in the Dutch West Indies often had their left arms bound. Hertz incorrectly assumed that in the West physical restraints, such as hand binding, had been abandoned, but he found the same aims in subtle, but no less insistent, practices: "We have abolished material bonds—but that is all." But the persistence of prejudice against left-handed children remained. "One of the signs which distinguish a well brought-up child," wrote Hertz, "is that its left hand has become incapable of any independent action" (92).

SOCIOLOGY, SOCIALISM, AND DREYFUS

Politics and ideology also played an important role in Hertz's formulations. In 1901, the 19-year-old Hertz had enrolled at the École Normale Supérieure in Paris, becoming a doctoral student of the famous French sociologist Émile Durkheim (1858–1917). Hertz was admitted to Durkheim's inner circle, including membership in *Année Sociologique*, Durkheim's journal and working group—which included such future luminaries as Marcel Mauss (1872–1950) and Maurice Halbwacks (1877–1945). Hertz formed a close friendship with Durkheim's nephew and protégé, sociologist and anthropologist Mauss, who was then researching the origins of magic, the function of sacrifice, and the creation of classification systems. In 1901 Mauss was appointed to the chair of the "History of Religion and Uncivilized Peoples" at the École Pratique des Hautes Études. Like Durkheim and Mauss, Hertz was Jewish and very much caught up in the debates about the 1894 conviction of Jewish Captain Alfred Dreyfus for spying for Germany. The Dreyfus Affair had launched pent-up anti-Semitism in France and, according to his biographer Robert Parkin, had persuaded the young Hertz to abandon his interest in becoming a politician in favor of researching the cultural origins of discrimination.[22]

Like his colleagues and mentors, Hertz supported the socialist, pacifist politician and journalist Jean Jaurès (1859–1914), a leading supporter of Dreyfus's innocence. But Hertz's socialism was also formed by his experience in Britain, where he conducted research at the British Museum from October 1904 to July 1905 and again in July and August 1906. During this period Hertz became enamored with British Fabian socialism, which advocated activism aimed at practical solutions to social problems. In particular he was attracted to the Fabian emphasis on inductive reasoning, in which social reality was determined through the collection of facts rather than from the assumptions of theory.[23]

Hertz was committed to social activism throughout his brief career. Thus, despite his support of the pacifist Jaurès, he eagerly volunteered for the French army in 1914—and requested an assignment at the front—when he could have easily claimed exemption because of his age and prior military service. The 33-year-old Hertz was killed in battle on April 13, 1915. Shortly after reenlisting, Hertz had explained his motives in a letter to his wife, Alice. "My dear, as a Jew, I feel the time has come to give *a little more than is due* [because] . . . there can never be enough Jewish devotion to this war, never too much Jewish blood spilt on French earth."* The second motive was political. He was a socialist, he wrote to Alice, and he wanted to demonstrate that community service could be as powerful a human motive as self-interest. Finally, Hertz wrote that as a sociologist he believed that citizens could be motivated to self-sacrifice by adherence to "the common good."[24]

DEATH AND THE LEFT HAND

These values, all of which affirmed Hertz's commitment to activism, cannot be ignored when examining Hertz's two published essays, both now regarded as anthropological classics. The first, "A Contribution to the Study of the Collective Representation of Death," published in 1907 in *Année Sociologique,* examines mourning practices in traditional societies, concluding that death was a social rather than purely biological event. For most humans, wrote Hertz, death signaled a "temporary exclusion" from society. The ritual segregation of the dead body enables a transition from "the living into the invisible society of the dead."[25] The completion of mourning defeats biological termination of life as the dead are reintegrated into the society as revered souls of ancestors. What was true of "primitive" culture could, according to Hertz, be found in contemporary society, though the process was less obvious to the untrained observer. For our purposes, Hertz's essay on death illustrates his larger project. That is, what appeared to be biological processes, like death and, as we shall see, human handedness, was equally determined by culture. If these cultural features, based as they often were on superstition and magical thinking, resulted in discrimination—for example, bias against left-handers—these features should and could be altered.

---

* Hertz was born in France, the son of a German immigrant father and a British American mother.

The Ambidexterity Culture Society

Hertz's interest in left-handedness had been aroused by his exposure to the British Ambidexterity Culture Society when he conducted research for his thesis at the British Museum in London from 1904 to 1906.[26] The movement's roots can be traced to the writings of British novelist and essayist Charles Reade in the late 1880s.[27] Despite Reade's efforts, the ambidextrous movement did not gain traction until the beginning of the twentieth century, with the publication of British educator John Jackson's widely popular book, *Ambidexterity: or, Two-Handedness and Two-Brainedness*.[28] In his preface to *Ambidexterity*, Sir Robert Baden-Powell, the founder of the Boy Scouts, wrote that if children were trained to use both hands equally both hemispheres would be stimulated, and the result would be more symmetrical, intelligent, and gifted children.[29] "It is time that all this sentimental vapouring about the imaginary advantages of Dextral pre-eminence and left-brain predominance should cease," wrote Jackson. Ambidexterity, he insisted, was natural in that it provided "each hand an equal chance in the race for supremacy." Encouraging symmetrical brain organization, ambidexterity results in greater efficiency for individuals and nations.[30]

Throughout Hertz's article on the preeminence of the right hand, Jackson's writings served as one of two main sources for his discussion of the treatment of left-handers and as the basis for evidence of the advantages of ambidexterity for both right- and left-handers.[31] The second source was Sir Daniel Wilson's 1891 book *The Right Hand: Left Handedness*.[32] Wilson (1816–1892) was president of the University of Toronto and a renowned scholar of prehistory. An enthusiastic advocate of ambidexterity, Wilson made two arguments that Hertz would later adopt in his essay. The first was that although handedness had organic origins, these were not a sufficient reason to assume that right- or left-handedness was immutable in individuals. Conceding that "true" left-handers could not be switched, Wilson noted that the retraining of many other left-handers demonstrated that both hands could be enabled (207). Pointing to athletes, carpenters, lumberjacks, painters, pianists, sculptors, dentists, and surgeons, Wilson touted the advantages of engagement of both hands and, therefore, both cerebral hemispheres. For Wilson, creativity and ambidexterity were intimately connected. Thus, Wilson believed, parents and teachers had a duty to encourage the use of both hands (205).

Hertz's conclusion to "The Pre-Eminence of the Right Hand" not only makes arguments that were indistinguishable from those of Wilson and Jackson but also adopts a tone similar to Jackson's. Acknowledging his debt to Jackson, Hertz asserted that the solution to the ancient prejudice against left-handers lay in the encouragement of ambidexterity: "Even if the different attributes of the two hands . . . are in great part the work of the human will, the dream of humanity gifted with two 'right hands' is not visionary." Indeed, Hertz was optimistic that the decline of religious dogmas created a unique opportunity to overthrow the predominance of the right hand: "If the constraint of a mystical ideal has for centuries been able to make man a unilateral being, physiologically mutilated, a liberated and foresighted community will strive to develop better the energies dormant in our left side and in our right cerebral hemisphere, and to assure by an appropriate training a more harmonious development of the organism."[33]

An important side effect would be the decline of anti-Semitism.

HERTZ AND THE JEWISH QUESTION

As noted above, in his letter to Alice, Hertz laid out three reasons for his volunteering to serve in the Great War. First, Hertz believed that his actions would demonstrate, contrary to the anti-Semitism that permeated French society, that the behaviors ascribed to Jews were not "racially" (biologically) determined. This would be evident from his and other French Jews' spilling their blood defending French soil. His socialist Fabianism was evident in his commitment not only to identifying social issues but also to liberating those, like left-handers (and Jews), from the irrationality that anthropology exposed. Finally, and consistent with Durkheim's sociology, Hertz was convinced a rationally ordered, inclusionary society would result in self-sacrifice and social cohesion.[34]

THE LEGACY OF LOMBROSO AND HERTZ

The contrasting views of Lombroso and Hertz continued to frame the debate of the etiology and effect of left-handedness for the next century, especially the controversy over whether to encourage or discourage left-handedness, which I elaborate in chapter 4. Although both men had claimed the authority of science, they disagreed on what constituted science and how it should be practiced. Partly this reflected their disciplinary differences, which influenced what sources they relied upon. Lombroso, trained in medicine, particularly what today we might label neuropsychiatry, grounded his arguments in neurology, genetics (of his day), and evolutionary theory. Thus he drew on constitutional medicine, modified Darwinian evolution, and degeneration theory. While Lombroso had idiosyncratic interpretations of all of these theories, they nevertheless provided the soil necessary for the growth of his claims that left-handedness was a pathological condition.

If Lombroso represented the neuropsychiatry of his day, Hertz reflected its rejection. Hertz's views had been nourished by the group of young scholars who in Durkheim's *Année Sociologique* took direct aim at the claims of physicians like Lombroso. Hertz's study did not engage the growing neurological investigations of laterality and hemispheric functions except insofar as he found them disputed in the proambidextrous writings of Wilson and Jackson. Hertz relied on anthropological and sociological studies (in this case, of aboriginal peoples) in a much more systematic way than Lombroso had. Yet, Hertz's understanding of left-handers came from books, archives, and the polemics of the Ambidexterity Culture Society, not fieldwork. Lombroso measured the bodies of dead criminals and the tattoos of living criminals but not a representative sample of left-handers.

Nevertheless, the issues that we confront today about the etiology, classification, extent, and consequences of left-handedness can be found in articles on the left hand written by these two early twentieth-century figures. The closer we come to our own era, the less obvious it may seem that, as Hertz claimed, our views on handedness reflect unexamined values rather than objective findings.

Most people living today believe, as both Lombroso and Hertz did in their time, that our current science provides the objective tools to evaluate data on handedness. Rejecting Lombroso's classificatory system based on morphological differences and measurements of the body, those who came after Lombroso believed that their classification systems, behavioral theories, and more sophisticated measurement tools provided objective evidence for their claims about the etiology and impact of left-handedness. Hertz, while acknowledging the biology of laterality, challenged the classification system of his day—and implicitly of ours—and suggested that elaborate subterranean beliefs underlay putative scientific objectivity. We do not have to go back a century, but only a few decades, to discover claims that from our current vantage point seem quaint and scientifically primitive. Only if we were self-deceptive would we conclude that our scientific tool set—high-powered biostatistical methods, sophisticated genetic models, state-of-the-art functional brain imagining, and the test for statistical accuracy, the p value—will make it possible for us once and for all to explain the riddle of human handedness. But acknowledging that our scientific claims and methods reflect unexamined values should not discourage us from trying to uncover the riddles of handedness or, as important, to recognize and build on the contributions of past and contemporary researchers who have attempted to solve its mysteries.

The consequences of classifying populations by biological traits that often are labeled as "race" continue to confront us. Like Hertz, Lombroso was distressed by late nineteenth-century anti-Semitism. When it came to explaining criminal behavior among Jews, Lombroso jettisoned his atavistic and biological arguments in favor of a cultural explanation. "His own Jewish ancestry," write Gibson and Rafter, "partially accounts for his refusal to characterize Jewish behavior in simple biological rather than more complex sociological terms."[35]

Lombroso's position was very close to Hertz's, that however much or little they were influenced by biological factors, it was anti-Semitism that restricted Jewish social interactions. However, Hertz's anthropology was consistent in its attempt to eliminate discrimination, whereas Lombroso's commitment to atavistic explanations for criminality and left-handedness forced him to make an exception when it came to the Jews, which others would refuse to endorse. Rather, others would include Jews among natural-born primitives, constitutionally incapable of adapting to modern civilized rules of behavior. Like Lombroso, they would justify discrimination on the basis of physical features that putatively could be observed and measured. Once the "type" was identified, they too would argue, the "type" incapable of being (re)integrated should instead be separated from the rest of society. For Lombroso the logic of his theory pointed to humanitarian solutions. He argued that the state should provide for the care of atavistic individuals because their behaviors resulted from no fault of their own. Others would conclude, for similar reasons, that the state should eliminate undesired types. Discrimination, whoever it is against, arises from the same soil, nourished by a belief system that is fallacious. For Hertz, eliminating those who were different (whether literally or by sidelining them and their contributions) impoverished not only the excluded individuals but also society itself. Society clearly has not taken up Hertz's beliefs except in small measure at a local level, here and there.

Lombroso's claims, albeit not his methodology or assumptions, have proven persistent. His connection of left-sidedness with feeble-mindedness and mental illness and left-handedness with criminality has had great resilience as it has been repackaged in contemporary and current scientific discourses. As Lombroso noted, negative attitudes toward left-handers reflect ancient prejudices about left-handedness. Indeed, in his work he was authenticating these ancient beliefs through "modern" scientific analysis. As he wrote, "long before I, after much technical observation, came to this conclusion, the people in the provinces of Emilia, Lombardy, and Tuscany had already declared the same when they framed and used the saying 'He is left-handed,' to express the idea that a person is untrustworthy."[36]

Without meaning to endorse this ancient stigma, Lombroso was correct in pointing out the persistence of the connection between cognitive deficits and left-handedness. But the question remains whether the deficits arise as a result of left-handedness, or as Hertz insisted, as a result of the stigmatization and practices aimed at left-handers.

But do we even know whom to count as a left-hander? Chapter 3 examines this issue.

3

BY THE NUMBERS
Measuring Handedness

The measurement issue hinges partly on whether handedness is fundamentally a matter of skill or a matter of preference.

M. CORBALLIS (1997)

Differences in the prevalence of left-handedness might be attributed to the divergent incidence of left-hand preference in divergent populations, divergences between populations in strength of cultural and environmental pressure against left-hand use, discrepancy of methods used for assessment of handedness or type of a criterion used for categorization of handedness.

Y. P. ZVEREV (2006)

Although I have always considered myself and my mother to be left-handed, depending on *what* is measured and *how*, each of us could be categorized as right-handed. According to the most widely used instrument, the 1971 Oldfield/Edinburgh Handedness Inventory, my mother would be right-handed because she performed most of the tasks in the survey, including writing, throwing, and sewing, with her right hand, even though her natural preference, until forcibly switched, was to perform all of these with her left hand. I write, eat, and throw with my left hand, but I bat baseball and play golf with my right hand. When I was younger, I could and often did eat with my right hand and play racket sports with either hand, but I was more skilled with my left hand. As I grew older I found myself becoming more left-handed. In my case it could be possible that I was genetically right-handed, but my training allowed or encouraged me to rely on my left hand, while my mother's resulted in her being categorized as right-handed.

Like my mother many "natural" left-handers were trained or succumbed to cultural pressures to be right-handed. That is, they developed the requisite skills to be counted as right-handers. As we will discuss at length in the following chapter, many normally left-handed children were and continue to be forcibly switched to right-handedness. Thus, whether a person is counted as left- or right-handed may depend upon whether genes or performance is measured.*

The issue in this chapter is not whether left-handedness is inherited but rather whether researchers are measuring socially adopted or hypothesized genetic factors in order to determine the extent of left- and right-handedness. However, there is no gene or reliable test available to determine handedness in utero or in DNA. Since children do not adopt a dominant hand until approximately 3 years of age, researchers are forced to determine the distribution of handedness from children (over 3) and adults, some or many of whom may have "chosen" their dominant hand despite their genetic inheritance. As we will see, this influences profoundly the determination of the distribution of left-handedness.

The instruments and methods used to assess handedness pose a major problem for determining the prevalence of left-handedness, challenging the commonly reported 10 to 12 percent prevalence of left-handers. New Zealand psychologist Michael Corballis writes that there is "little agreement as to precisely how handedness should be defined or measured" even though "nearly all people readily identify themselves as being either right-handed or left-handed."[1]

The assumption is that *Homo sapiens* have been 90 percent right-handed since the late Stone Age,[2] but there are great disparities in the reported prevalence of left-handedness in different populations and cultures. It is very difficult to pin down exact percentages of left-handers and right-handers. Still, the reported rate of left-handers tracks with the prevalence of discrimination against left-handers in any society: the lower the reported prevalence of left-handers in a society, the greater the stigma and discrimination toward left-handers.

---

*The number of left- and right-handers depends on what is measured. In the handedness literature a distinction is made between "preference" (presumed genes) and "skill" (task performance). I find the distinction between preference and skill confusing, especially the use of term "preference" to denote something that is biologically determined. Thus in what follows I distinguish between whether the rates of left- and right-handedness are determined by measuring predicted putative genetic causes, from performance, or some combination of the two.

MEASURING HANDEDNESS

Assessment of left-handedness is generally based on one or two of three criteria: observation, self-reporting, and/or a survey questionnaire, administered either by an observer or filled out by subjects themselves. Of the three, observation alone, with its well-recognized observer bias, is the least reliable. Self-reporting has the advantage of allowing the collection of large samples, but, similar to observer reports, suffers from well documented unreliability.[3] The third and most reliable method, because all respondents answer the same questions, is a survey inventory.[4] A number of survey inventories are designed to identify handedness, most of them brief and administered easily and quickly. The most widely used in Britain and the United States is the Oldfield/Edinburgh Inventory, reproduced below, consisting of 10 nonweighted questions.[5] Other similar inventories include British psychologist Marion Annett's 12-question survey.[6] Although these surveys are reliable, in that they measure the same things, they may not be valid.[7] For instance, both the Annett and Oldfield inventories weigh all their variables—including broom sweeping and matchbook lighting—as equal in value to eating and writing.[8] Moreover, the

	Left	Right
1. Writing		
2. Drawing		
3. Throwing		
4. Scissors		
5. Toothbrush		
6. Knife (without fork)		
7. Spoon		
8. Broom (upper hand)		
9. Striking match (match)		
10. Opening box (lid)		
TOTAL (count Xs in both columns)		

The Oldfield/Edinburgh Handedness Inventory, the most widely used instrument to determine handedness in Britain and the United States, consisting of 10 nonweighted questions.

questionnaire asks about activities, such as sweeping and shoveling, that many subjects do not do.[9]

To his credit, Oldfield acknowledged the limitations of his inventory. Its advantages are that it is simple to administer and that it "provides one quantitative measure of handedness backed by a known distribution of values in a reasonable sized normal population."[10] Oldfield's inventory was designed to measure the extent of handedness in populations rather than as a test for individual handedness. As Oldfield explained, it is most useful for providing a standard for comparing large populations (110). "In the absence of any firmly based knowledge of the underlying mechanism of handedness," he wrote, "the only way of providing such a measure is to adopt a set of inventory items and a scoring and computational convention, and apply these to an adequate sample of individuals."[11]

DISPARITIES: REAL OR CONSTRUCTED?

A 1980s survey of more than 20,000 mainland Chinese students and professionals reported that only 0.23 percent were left-handed, an astoundingly low number.[12] Thirty years later, the numbers hadn't changed much.[13] Depending on the definition used and the geographic location, the prevalence of left-handers ranges from 0.06 to 25 percent.[14] A meta-analysis of throwing in 14 countries (1.2 million persons) for the period 1922 to 1998 reported a prevalence of left-handers between 5 and 25 percent. In general the highest rates (15% to 20%) are reported in the United States, Canada, Britain, and Western Europe, with the lowest (0.06% to 2.8%) in the Congo, China (including Hong Kong), and Taiwan. The overall finding is that the prevalence of left-handers is significantly lower in Asia and Africa than in the West and (nonaboriginal) Australia.[15] Whether these disparities reflect biological or cultural differences remains contested.[16]

The disparities in estimated size of the left-handed population exist not only between countries but also within the same country. For instance, the estimated percentage of left-handed US male throwers ranged from 7 to 9.4 percent, while a study of the percentage of left-handed hammerers in France ranged from 15 to 21 percent.[17] Similarly, a 2006 British (BBC) Internet study of 250,000 participants (half male, half female) from seven British ethnic groups reported that between 7 and 11 percent of the participants preferred their left hand for writing, with the highest percentages among those who self-reported being "white."[18]

Such disparities point to a potentially large margin of error in the reports of the frequency of left-handers, which makes developing robust statistical claims about the prevalence of left-handedness problematic. As Malawian University physiologist Y. P. Zverev writes, these differences reflect a "discrepancy of methods used for the assessment of handedness or type of a criterion used for categorization of handedness."[19] Unlike in developed countries, where handedness is generally measured using the writing hand, studies of traditional cultures have relied on eating, throwing, or tool use. Studies that used throwing or hammering to determine who was left-handed reported twice as many left-handers as those that relied on writing.[20] The implication seems to be that for acts less likely to come under social-cultural control and/or to be the target of special training, left-handedness is more prevalent than it is for acts, prominently eating and writing, more likely to come under social-cultural control. In sum, there is no consensus on the prevalence of left-handedness, in large part because there is no agreement on what is to be measured.

Although males are more likely to be left-handed than females, the ratio remains unclear if consistently uneven.[21] While most studies indicate that for every 5 male left-handers there are only 4 females,[22] others report the ratio to be as high as 2:1.[23] But whether the ratio is 2:1 or 5:4, sex differences were strongly confirmed in a recent meta-analysis of 1.8 million subjects.[24]

What accounts for these sex differences? And why is it that in almost every culture the left has been associated with female and the right male, but all the data demonstrate that females are much less likely to be left-handed than males? Is the reported difference between males and females an artifact of how left-handers are identified? We know that while male left-handers *may* outnumber female left-handers, that ratio (or its inverse) does not necessarily align with the ratio of males to females with right-hemispheric language.

A number of studies have concluded that language is more bilateral and diffuse in females than in males.[25] In addition fMRI evidence suggests structural differences between males and females. In particular, the planum temporale (the upper surface of the temporal lobe that overlaps with Wernicke's area) appears to be more pronounced in males than females.[26] Thus there seems to be both functional and structural differences in language lateralization between males and females. However, these findings do not explain why females are less likely to be left-handed than males. Indeed, as we will learn throughout this book, reduced lateralization has been associated with increased incidence of left-handedness and learning disabilities. If the language laterality data on sex are valid, one would predict that females would be more likely than males to be left-handed. But the opposite seems to be true.

Although there are a number of comprehensive studies of left-handedness among Africans, as I noted in chapter 1, there have been surprisingly few investigations of left-handedness among African Americans. Those that exist are contradictory. For instance, a team of University of California psychologists reported in 1975 that behavioral testing of approximately 8,000 elementary students, grades 1 through 6, found that race had "little meaningful relationship to the incidence of handedness."[27] Subsequent investigations, however, have reported a greater prevalence of left- and mixed-handed African American males than among their European American counterparts. In 1985 D. A. Saunders and A. L. Campbell, using "a modified version of the Edinburgh Handedness Inventory," examined "the degree of left-, mixed-, and right-handedness" among Caribbean and African American students attending Howard University. They found that African American men had a significantly higher prevalence of left- and mixed-handedness than African American women. Moreover, left- and mixed- handedness were "significantly higher" in African American males than "reported in some other studies on Anglo-Saxon populations." Saunders and Campbell find "the high incidence of left-handed men in this study is striking, in light of anecdotal evidence that cultural biases still exist against sinistrality in Caribbean societies and, in some cases, among the grandparents of the USA subjects."[28]

Examining 2,083 randomly selected adults, University of Cincinnati researchers in 1988 confirmed the Howard University data. They reported a prevalence of 9 percent left- and left-mixed among African Americans, compared to 6.7 percent of European Americans. Moreover, they found that the ratio of African American left- and mixed-left-handed males exceeded that reported in previous investigations. In addition, they discovered that younger African American males tended to be more left- and mixed-left-handed than their older cohort. The authors hypothesized that "different responses to cultural pressures might account for these race differences." These included the possibility that "for young black males, mixed-handedness is lauded." Another factor was that African American families place "less pressure on all children to conform to the right-handed norms of the dominant white society. Thus children with any tendencies to left-handedness receive relatively few clues that 'Right is right.'"[29] Taken together these few studies are suggestive but hardly definitive. Given the fact that the most recent of these investigations is three decades old, it is time for a comprehensive investigation into handedness among African Americans.

Along with the data in the studies of African American handedness, a number of investigators have reported that left-handedness declines as a person ages, which was true of both my mother and me. Stanley Coren and his colleagues reported that while 15 percent of 10-year-olds were left-handed, only 5 percent of 80-year-olds were.[30] This trend has been reported even among secondary schoolchildren. For instance, a French–Côte d'Ivoire team developed a 20-question manual preference test to determine the handedness of 382 secondary school students (ages 12–22) in Abidjan, the largest city of the Côte d'Ivoire. They found a 14 percent prevalence of left-handers at the younger end of the age spectrum (12- to 15-year-olds) that declined dramatically to 1 percent among 18- to 22-year-olds (151, 155–157). A second study by the same research group, but using a different 25-item questionnaire, of 759 Khartoum (Sudan) 18- to 33-year-old undergraduates, reported a 5 percent "left manual preference."[31]

This claim, like almost every other claim about left-handedness, was contradicted by other studies. Thus, a French team's survey of left-handedness in Algeria, Greece, Italy, France, and Spain found no decline of left-handedness with age.[32]

Coren and colleagues nevertheless insisted that the age-based decline in the prevalence of left-handers was real and reflected "reduced longevity" of left-handers.[33] This claim, though widely publicized and cited, has been overwhelmingly rejected by most other researchers.[34] Coren and his critics are probably using different definitions of left-handedness, which often seems to be the case for those seeking to uncover the prevalence of left-handedness. Nevertheless, even if the reported data are often unreliable, the reported rates of left-handers serve as an indication of the extent of stigmatization of left-handers. This is evident in the reported prevalence of left-handers in China and India.

IDENTIFYING LEFT-HANDEDNESS

As noted, reported prevalence of left-handedness is strongly influenced by how handedness is defined. In China, the reported prevalence of handedness was obtained by measuring the handed activities that were most essential to the Chinese when they were being measured. In earlier epochs, what mattered most was what hand one used for eating.[35] By the late twentieth century, writing and drawing had become the test for handedness. As University of Hong Kong researcher R. Hoosain pointed out in 1990, because natural left-handers were routinely made to write and draw with their right hands, little attention was paid to these people's continued reliance on their left hand for other tasks.[36]

What is found in China is found in other societies. Maria De Agostini and her colleagues asked their Côte d'Ivoire and Sudan subjects to identify pressures to change their writing, eating, or other handed activities. The team reported that in both societies the greatest pressures aimed to change left-handed eaters.[37]

Despite the low prevalence rates reported in the world's two most populous societies, China and India, most current experts claim that between 10 and 12 percent of humans are left handed.[38] For instance, psychologist and neurosurgeon Chris McManus cites an unpublished 1996 paper reporting the results of a 10-question survey of 2,892 North London schoolchildren aged 6 to 15 to determine the extent of their left-handedness.* Almost 10 percent were left-handed; 11.6 percent of males compared to 8.6 percent of females.[39]

---

* The children were asked with which hand they performed the following tasks: Write, Draw, Throw a ball, Brush your teeth, Hold scissors, Hold a knife (without a fork), Hold a Spoon, Hold a cup, Use a TV remote, Open a can of fizzy drink (with a ring pull). A score of five or more tasks done with the left hand would classify a child as left-handed.

Does Ambidexterity Exist?

So, what about the so-called ambidextrous? Not surprisingly, there is no agreement on what constitutes ambidexterity, with some researchers concluding there is no such thing.[40] Others restrict the term to those who randomly perform tasks with either hand.[41] Most commonly, however, the designation has been applied to those who use the one hand for certain tasks, like eating, while relying on the other hand for writing or throwing. Thus the prevalence of the ambidextrous, based as it is on different criteria by different researchers, is impossible to determine, which does not mean that researchers ignore it. According to a recent study, many of the so-called ambidextrous studied were left-handers who had been made to switch for certain tasks and not others.[42] In many cases the retraining took place in early childhood, although the subject may not recall the training. Thus, some researchers argue that the people who initially were left-handers are not truly ambidextrous, though how this population can be easily identified remains challenging.[43] Sometimes these subjects appear to be indecisive handers, labeled as non-right-handers,[44] a classification that some experts reject.[45]

As with ambidexterity, there are no agreed upon definitions of non-right-handedness. The designation non-right-hander is not synonymous with the categories of strong or weak left- or right-hander as indicated by an evaluation by an instrument such as the Oldfield or the Annett inventories. Rather, it is a device called upon to justify a conclusion when there is no statistically significant finding of left-handedness based on standard inventory classification. It generally, but not always, indicates a right-hander who does not exclusively rely on the right hand, right foot, right eye, or right ear but who does not fit any classification as a left-hander. For some researchers, non-right-handedness includes "nonconsistent right-handers," in contrast to "exclusive right-handers" and subjects "who tended to favor the left hand versus participants who tended to favor the right hand."[46]

"My least favourite term," writes McManus, "is 'non-right hander,' a term particularly in vogue in the 1970s and 1980s, which seems to have been popular because it appears to make fewer judgements about what was and was not a left-hander." But, notes McManus, "in practice it seems to solve nothing but merely provides the user with a veneer of pseudo-scientific precision." This is no trivial issue, because "almost any meaning can be attached to it, and there are several studies in which anyone who does not score zero on a questionnaire . . . is described as a 'non-right-hander.'" As important, writes McManus, is "the serious practical problem that it gives the incompetent or unscrupulous researcher a choice of a wealth of measures, some or other of which may attain statistical significance due to chance alone."[47] Despite these reservations, as we will see, the designation is frequently invoked by those who insist there is an increased risk of learning disabilities among those with putatively "abnormal" handedness.[48]

PREVALENCE AND DISCRIMINATION

As the example of the aboriginal Arunta of Australia discussed below suggests, reported rates of left-handers that are significantly lower than 10 percent suggest sustained discrimination against left-handers.[49] Some natural left-handers may, through either shame or fear, underreport their left-handedness. Others may successfully pass as right-handers as they voluntarily shift to right-hand reliance. Other natural left-handers may have been forced to switch to using their right hands. To the extent that handedness has a genetic substrate, the rejection of left-handers as marriage partners would result in a population decrease of left-handers. From these perspectives, contemporary China, with its incredibly low prevalence, would represent a high level of discrimination, whereby the United States, Canada, and Western Europe would represent a high level of tolerance.

Practices resulting in the reduction of left-handers in a population have been referred to as "the cultural pressure hypothesis." Psychologist John L. Dawson found support for the hypothesis in his comparison of the Temne of Sierra Leone with Arunta aboriginals of Central Australia. The Temne culture, being more conformist, reported a 3.4 percent left-handed rate, compared with the 10.5 percent reported among the more permissive Arunta.[50] This meshes with Zverev's assertion that in contrast to Western societies, in many African societies, like the Temne, powerful cultural pressures suppress the expression of left-handedness.[51]

The methods used to persuade left-handers to switch are not always negative. For instance, Bryden and colleagues reported that in the Amazon region of Colombia, positive reinforcement is used to increase right-hand preference among Tucano adolescents.[52] By rewarding right-handed use beginning in early childhood, the Tucano have been more successful than other cultures in transforming left-handed children to right-handers in all activities.[53]

Discussing handedness in "China" or "India" (or even more problematically, "Africa") requires painting with a very broad brush. These diverse and complicated societies ultimately must be examined in more detail with the recognition that even the most careful and robust local studies should be generalized only with the greatest of caution. Thus in India or China, not to mention the diverse African continent, the prevalence of left-handedness can vary regionally and within the same regions, by ethnicity and religious practice. For example, Krishna R. Dronamraju found a high left-handed prevalence among Andhra Pradesh tribals compared to Hindu men and women in Andhra Pradesh.[54] Whether or not Dronamraju's data have been borne out, his observations are reinforced by the recent growth of web-based sites such as The Association for Left-Handers in Mumbai calling for the cessation of the practice of forcing left-handers to convert to right-handedness.[55]

Both traditional values and the beginnings of modernization can depress the prevalence of left-handers. The role of traditional factors is widely acknowledged, and researchers have examined these factors extensively. They have also produced a number of studies concluding that the centralization of resources that drives urbanization might have a profound impact on rates of left-handedness. This effect is especially visible in educational institutions where the influx of a first generation of children to be educated has strained resources and required sustained conformity so students can be taught basic skills, especially to read and write. From this perspective, allowing 10 percent of the students to write with their left hands has been seen as wasteful and inefficient, especially in societies that already have a deep antipathy toward left-handedness. This is reflected in data from China, Africa, and South Asia showing that as students become more integrated into educational institutions, their prevalence of left-handedness declines. Thus, secondary education students became more right-handed than their younger siblings in elementary grades. As the developing world emerges into a collection of densely populated urban centers and as rural migrants become integrated into modern society, diversity, including encouraging 10 percent of the population to develop their left-handed skills, is inefficient.

Recent studies suggest that a similar scenario, of increased public education resulting in a decrease in left-handers, took place in nineteenth-century United States, Canada, Britain, and Australia. University of Melbourne researcher C. J. Brackenridge reported a 2 to 13 percent increase in high economic status among left-handed writers born in Australia or New Zealand between 1880 and 1969. Brackenridge attributed the increase in left-handers to what he described as "cultural relaxation" toward left-handers. He suggested that a similar increase in left-handers would also be found in the United States.[56] A comparable pattern was found in the Netherlands, where researchers revealed that almost no left-handers 40 years or older used their left hand for writing, while nearly 100 percent of left-handers 14 years old or younger did.[57] The authors attributed these results to changing attitudes toward left-handers by Dutch educators.

Similar conclusions are supported by data that connected the early twentieth-century debate among American and British educators over switching (discussed in chapter 5) to increased tolerance toward left-handed writers.[58] This was affirmed by American statisticians Avery N. Gilbert and Charles J. Wysocki, who analyzed data on left-handed writing and throwing that were collected as part of the 1986 *National Geographic* "Smell Survey" of 1.78 million American men and women.[59] Employing sophisticated statistical methods and a novel set of variable phenotypes, Gilbert and Wysocki reported findings supporting the hypothesis that as social discrimination against left-handed writers declined, the reported prevalence of left-handers gradually increased.[60] Based on this hypothesis, the repression of left-handed writing in the United States would appear to have eased first around 1917 and been fully lifted by 1937.[61]

The implications of this finding for McManus and his colleagues are these: in late nineteenth-century North America and Britain there was a decline in left-handers followed by a significant increase in prevalence in the twentieth century. Combining Gilbert and Wysocki's data with a number of other studies, including a large 1953 data set collected from 6,549 BBC viewers, the McManus team concluded that left-handedness reached its lowest prevalence in the 1890s, declining from 10 percent at the end of the eighteenth century to 7 percent in the 1830s, and on to its lowest point, 3 percent, at the end of the nineteenth century. After that the prevalence began to increase rapidly, reaching its present rate following the Second World War.[62]

Here and in other publications, McManus insists that the decline in left-handers represents a decline of what we might call "naturally born left-handers" rather than a reflection of left-handers passing themselves off as right-handers or having been switched to right-handedness by nineteenth-century parents, teachers, and other accumulated social forces. While this may be so, most of the data about nineteenth-century Anglo-American left-handedness are based on self-reporting. Given the strong social stigma toward left-handedness in the nineteenth century, many natural left-handers may have purposely misrepresented their handedness; others may have forgotten the training that directed them to switch. In either case, many of those switched to right-handed writing would most likely see themselves as right-handed if they used their right hand for writing. Also, to the extent that Coren and others are correct in claiming shorter life spans for left-handers,[63] many of the older left-handers may have died by the time of the surveys, explaining why there were so few older left-handers, especially in the Gilbert and Wysocki data. It is possible that as people aged they tended to use their right hand more for writing and throwing—that this population had been under greater pressure to use their right hand than previous or subsequent populations. But, McManus could be correct in asserting that there were fewer natural left-handers and, as we shall examine later, he has proposed a genetic model that supports his conclusion that discrimination against left-handers made them less desirable mates. As a result, those carrying the putative gene that makes left-handedness possible often married later or less frequently and had fewer offspring, thus reducing—but, as we will see, not eliminating—the transmission of the hypothetical gene responsible for left-handedness.[64] (The genetic component of handedness is discussed in chapter 8.)

Whether the change in numbers of left-handers was real or merely reported, the fact remains that a reduction of numbers of left-handers resulted in great measure from switching left-handed writers to right-handed writers. Finally, especially in North America, the period of the supposed decline of left-handers maps on to the period of greatest immigration. As a number of studies have shown, immigrants, especially those who arrived from the 1880s to the First World War, who were typically from Southern and Eastern European cultures, would have been likely to restrict their children's left-handed behaviors and to be vulnerable to having others impose right-handed writing on them.

By the mid-twentieth century, the prevalence of left-handers increased dramatically in North America and Britain, and, soon after in much of Western Europe. A 1997 study of the handwriting habits of 700 French men and women between the ages of 16 and 80 reported that while only 9 percent of natural left-handers over 40 used their left hand for writing, 80 percent of those under 40 wrote with their left hand.[65] Similar findings have been reported for Germany, Italy, and Spain, suggesting that by the late twentieth century, educational practices had become more tolerant toward left-handers writing with their left hands.[66]

Although McManus and colleagues find it "indisputable" that the prevalence of left-handers declined in the nineteenth century, they admit to having "no robust evidence to suggest how and when" the increase noted in the early twentieth century occurred.[67] One possible source of evidence may be the resistance by educators and nascent child development researchers of the early twentieth century. This resistance came from two directions. The first can be seen in the proambidexterity movement of the late nineteenth and early twentieth centuries in North America and Britain, discussed in chapter 2. The second and more substantial reaction was the debate over the purported risks of learning and speech disorders, especially stuttering, that allegedly resulted from converting natural left-handers (explored fully in chapter 5).[68]

The answer to the question "Why are there no Chinese left-handers?" has a number of possibilities, including that there are and always have been many left-handers in China but that their low numbers are an indication of how they are measured. Alternatively, the differences noted between China and other countries in left-handed prevalence could result from a combination of traditional values and practical considerations that seemingly merged to reduce the *reported* prevalence of left-handedness. The question remains whether any of these factors reduced the *actual* prevalence of left-handers.

For many researchers left-handedness serves as proxy for other conditions and behaviors. Even if we were to find a way to address the measurement limitations of handedness inventories, we still would be faced with the even more complex issue of language lateralization. Indeed, Oldfield had warned at the outset that his inventory should not be used to measure cerebral laterality.[69] Nevertheless, inventories and other instruments created to measure handedness are employed regularly as proxies for hemispheric language laterality—and yet the vast majority of left-handers, contrary to what one might expect, are left-brained for language.

What are we measuring and what do we hope to explain by what we have measured? As we have seen in this chapter, the identification of left-handedness is problematic for a variety of reasons, not least of all, writes Corballis, because it "hinges partly on whether handedness is fundamentally a matter of skill [task performance] or a matter of preference [genetic inheritance]."[70] Although task performance and genetic pressures are different factors, in combination they can and most assuredly do both play a role in the determination of handedness. But, the problem remains that researchers have tended to rely on either one or the other rather than a combination of both. As we have seen, this has had a profound influence on what researchers measure and what they find.

4

Ambiguous Attitudes

I can attest to the life-long disadvantages that can accompany being forced to change hands. All of my life, I have had the sense that I have been forced to compete in the world with one hand tied behind my back, kind of like Kirk Douglas in the movie "Lonely Are The Brave" [1962] . . . [where] Douglas is forced to fight an amputee war veteran with one hand (his good hand) tied behind his back. That movie had a huge impact on me as a child and now I know why.

ROBERT WILLIAMS TO H. KUSHNER (JULY 18, 2016)

The superior man ordinarily considers the left hand the most honorable place, but in time of war the right hand, . . . To consider this desirable would be to delight in the slaughter of men and he who delights in the slaughter of men cannot get his will in the kingdom.

LAO TSE (600 BCE)

It is said amongst the old Zulus that no person of importance ever counted with his left hand.

DUDLEY KIDD (1906)

As a child I wondered why my mother had been "retrained" to use her right hand while I had not. I assumed that forcing left-handed children to become right-handers reflected some old-fashioned prejudice that had disappeared by the time I was growing up but might still be found in traditional societies. The practice was described in Dudley Kidd's 1906 ethnography of Zulu children who were forbidden to eat with their left hand. If a Zulu child persisted in eating with his or her left hand, the child's left hand was placed in boiling water. "By this means the left hand becomes so scalded that the child is bound to use the right hand."[1] Like the Zulu, many traditional North and East African peoples actively attempt to "cure" left-handedness.[2]

As I investigated switching left-handers more systematically, I was amazed to learn how widespread the practice was and continues to be globally. When someone learns I am writing a book about left-handers, they often volunteer that parents, grandparents, or close relatives were forced to switch their writing and eating to their right hands. Following my interviews in print or talk media, I invariably receive emails detailing the sender's own, never positive, hand-switching experience. Responding to my March 2016 interview on ABC Australia,[3] Michael Hackh, a 61-year-old German engineer from Wiesbaden, emailed me: "I'm left handed but was trained/educated to be a right hander, when I was a small child being 2 or 3 years old." Hackh is convinced "that such a conversion of handedness and training . . . leads to a wide area of psychological and health problems especially in the second half of the life." He is certain that "converted handedness" was responsible for "all kinds of sicknesses, but unfortunately I can't prove anything."[4] Along with other retrained German left-handers, Hackh set up an Internet forum, Forum für Linkshänder—Das deutsche Linkshänderforum, which attracts hundreds of visitors weekly. Many post their own experiences or that of a parent, invariably negative, about their retraining.[5] Similarly, American author Samuel M. Randolph details an array of disorders that he attributes to his retraining, including profound brain alterations that he claims were reversed by his relearning to be a left-hander at age 41.[6]

These responses, along with Randolph's book, have reinforced Hackh's belief "that about 50% of all humans on this planet are left handed, but about 80% of all left handers don't know that they are converted, due to education and social adaption processes beginning in the earliest years of small children." One does not have to agree with Hackh's impression about the percentage of natural left-handers or his belief about the effects of retraining on his health to recognize the profound impact that this practice has had on many people.

THE ZULU, BRITISH PRINCES, AND THE REST OF US

As Hackh's website demonstrates, forced switching of left-handers, known as "retraining," was not restricted to traditional societies. Nineteenth- and twentieth-century European and American physicians and educators advocated compelling children to eat, write, and perform other tasks with their right hands. Rejecting the notion of natural left-handedness, these experts, much like the Zulu elders, asserted that a child's decision to rely on his or her left hand was a reflection of a defiant personality that could best be corrected by retraining.

The methods used by moderns were as torturous as those employed by the Zulu. These included tying up a resistant child's left hand, which forced the child to rely entirely on his or her right hand.[7] Recalling his experiences in a mid-1920s elementary school, one British left-hander reported that he and others were forced to sit on their left hand and write with their right hand. When that failed, his teachers tied his hand behind his back. He was regularly humiliated. Teachers shouted at him to use his right hand, and when he resisted he was made to stand in the corner and forbidden to participate in classroom activities. "We were told we were hopeless and everyone in the class repeated it."[8] Another related that whenever he used his left hand his teacher poked a cane in his ribs. "He was very nasty about it and frightened me to death." Even as an adult the former pupil recalled this teacher with dread.[9]

"Retraining," as it became known, was not limited to the weak and powerless. The Duke of York, future King George VI (1895–1952), had been forced to write with his right hand.[10] The methods used to transform the Duke of York into a right-hander were similar to those reported by other British elementary students, including immobilizing his left hand by tying it down. Retraining was common in the West until the late twentieth century, as evidenced by an email from Dr. Robert Williams (a pseudonym). Williams, now a university professor, reports that his pediatrician father initiated retraining when the boy was 7, in 1964. Williams soon developed a mild stutter and a "diagnosis of hyperactivity" that resulted in treatment . . . by regular doses of Atarax, an early anxiety-reducing drug." Ever since his retraining, writes Williams, "I have had the sense that I have been forced to compete in the world with one hand tied behind my back." This practice should be stopped, implores Williams because "forcing kids to change hands [is] damaging both neurologically and psychologically."[11]

Nevertheless, many parochial schoolteachers continued to force left-handers to write with their right hands well into the 1970s.[12] "When I was in school," (in France) wrote 64-year-old Gisèle, "the nuns were teachers and they forbade me to write with my left hand because they said that the LEFT-HANDED were possessed of the devil." She was regularly hit with a ruler, and her hands were tied behind her to restrain use of her left hand. "In real life," wrote Gisèle, "everything is positive right and negative to the left." Born in Alsace in 1950 Beatrice recalls that she "was thwarted in a kindergarten" run by nuns. She was bullied and, like Gisèle, hit with a metal ruler to force her to cease using her left hand. It was, she writes, "a terrible year" in which Beatrice was "constantly terrorized." Subsequently she developed "a severe stutter" and "locked myself in a silence until the age of 11 years." She was also persecuted at home, where her "grandfather called me *left leg,* a stinging insult in Alsace." Later in grammar school her sewing teacher made fun of her and then "put me aside and ignored me throughout the year." Beatrice later fought back, and now as a teacher she has dedicated herself to helping other left-handers achieve their potential.[13]

Jean Pierre Zapata, born in 1948, wrote in 2015 that he grew up in the small village of Drôme. Left-handed and dyslexic, Jean was terrorized by an old male teacher, who "follow[ed] the guidelines to the letter: a lefty is sick! a dyslexic an idiot!" Jean "suffered all the bullying and possible punishments," including his "left arm tied behind his back." Nevertheless, he "liked school … [and I] was very inventive and gifted in many areas, but the school did not love me." "My trauma," insists Jean, "was real and I had to fight" to survive. He insists that because of his retraining he never learned to write or speak French properly. Retraining continued in French schools well into the 1960s, despite a 1947 report of the French Ministry of Education that urged abolition of the practice of forced hand switching.[14]

Reporting on their investigations of 650 British Columbia undergraduates, psychologist Clare Porac and colleagues found that attempts to switch left-handers into right-handers were common in Canada in the 1980s.[15] Combining the findings of two of Porac's studies, Coren found that 64 percent of left-handers recalled attempts by others to switch them to right-handedness prior to their ninth birthday.[16]

As journalist Natalie Jacewicz insists, bias against left-handers has not entirely disappeared in the West.[17] Retraining continues to be practiced in much of the non-Western world, including by recent immigrants to the United States, Britain, Germany, and France. Among the Islamic world's 1.8 billion people, left-handedness continues to be discouraged, and the forced transformation of left-handers into right-handers continues unabated.[18]

Why is retraining so enduring and widespread? When did the attitudes and beliefs that justify and sustain these practices arise?

Negative Attitudes

Negative attitudes toward left-handers have been deeply entrenched among humans for ages. The words *left* and *left hand* in almost all the world's languages are synonyms for "defective" or "sinister."[19] The word for left in Latin and Italian is *sinistra*, in French it is *gauche*, and in English *left* comes from the word *lyft* for broken, while in German, *linkisch* is associated with awkwardness.[20] Labeling someone as left-handed (*levja*) in Russian is a metaphor for deceptive or untrustworthy.[21] In Mandarin (Chinese) the character for left, *zoû*, is variously translated as weird, unorthodox, wrong, incorrect, different, contrary, or opposite. Taking the left path means using unorthodox or immoral means. In contrast, the Mandarin character for right indicates that one should eat with the right hand.*

Given the association of left-handedness with inferiority, it has unsurprisingly been characterized as feminine. Thus, regardless of data showing that males are more likely to be left-handed, left-handedness has been gendered female in most languages. In Chichewa, the traditional language for the majority of Malawians,† the word for left implies inferior, female, and weaker, while the right hand is often called the male hand.[22] This characterization of left-handers as unclean authorizes practices of cultural discrimination and social segregation, including restriction from performance of sacred duties.[23]

Stigmatizing left-handers, whether for their physical or supposed "social differences," has been reinforced in the religious texts of Judaism, Christianity, and Islam. According to the early medieval Persian scholar and historian Abu Ja'far Muhammad ibn Jarir al-Tabari (839–923), "Allah has nothing left-handed about him, since both his hands are right hands."[24]

---

* The Chinese use direction rather than left and right. Where in English we may say, "turn right at the crossroad," the Chinese will say "go westward," etc.

† Chichewa is a Bantu language, and along with English, it is the national language of Malawi.

The French anthropologist J. Chelhod reports that Islamic peoples have a strong preference for the right that is reflected in a variety of beliefs, practices, and restrictions.[25] Similar to other societies, in Islam, the right hand is associated with power and is the subject of extensive training, while the left hand is portrayed as synonymous with evil (242). In his comprehensive review of attitudes and practices of North and East African societies, anthropologist Heinrich Albert Wieschhoff (1906–1961) reported that the right hand was considered socially and morally superior to the left. Wieschhoff found that belief in the superiority of the right hand was strongest in Islamic cultures of North and East Africa.[26] The persistence of these cultural influences has been validated by Malawian University physiologist Y. P. Zverev in his studies of twenty-first-century Malawi students in Blantyre, the nation's second largest city. In Malawian society, writes Zverev, greeting someone or eating with the left hand are especially offensive acts—they are taboo.[27] He found that 87.6 percent of those surveyed believed that left-handers should be forced to use their right hand for habitual activities, including writing, tool use, and throwing. This belief reflected practices, for the study also found that all ambidextrous children and 90 percent of left-handers had been subjected to pressures to adopt right-handedness. When asked why they believed left-handers should be forced to switch to right-handedness, 38 percent responded that "the left hand is dirty, or it is disrespectful and unwelcome to use the left hand in public." A majority assumed that left-handers were less skilled than right-handers.[28]

Less stigmatizing but nevertheless negative attitudes toward the left are evident in Buddhism, where the road to Nirvana is on the right. In the combination of Buddhism and Hinduism known as Tantra, choosing the "right-hand path" represents traditional Hindu practice, including asceticism, while eating animal flesh, consuming alcohol, and certain sexual practices are associated with the left hand.* Even without a strictly religious justification, the Greeks and Romans, despite Julius Caesar's left-handedness, viewed the right hand as superior.[29] Bertrand points out that in the medieval church

---

* Journalist Melissa Roth notes that in Tantra both paths can lead to enlightenment but choosing the left is seen as involving greater risks and thus not advised for most followers (27).

the left was "synonymous with deviation, of depravation." He cites the fourteenth-century French Benedictine monk and poet Gilles Li Muisis, who wrote that good priests would be found among right-handers and that left-handers should use their right hands for carrying out sacraments.[30]

ON THE OTHER HAND

The negative attitudes toward left-handedness, while persistent, were not universal. In *The Republic* (380 BCE) Plato rejects the popular negative characterization of the inferiority of the left hand and condemns the practice of restraining its use. Conceding that the right and left hands have different tasks, Plato points out that, like the two feet, the two hands are complementary: "In natural ability, the two limbs are almost equally balanced; but we ourselves by habitually using them in a wrong way have made them different." He laments "that we have all become limping, so to say, in our hands," which Plato attributes "to the folly of nurses and mothers." There are great advantages to employing both hands, such as in the Scythian custom of nurturing the ability to draw a bow and arrow with either hand. In war, especially in close quarters, Plato asserts, reliance on both hands provides a deadly advantage. Pointing out that most successful athletes and warriors train both sides, Plato concludes that "in regard to the use of weapons of war and everything else, it ought to be considered the correct thing that the man who possesses two sets of limbs, fit both for offensive and defensive action, should, so far as possible, suffer neither of these to go unpracticed or untaught." In fact, "if a man were gifted by nature" to possess a "hundred hands he ought to be able to throw a hundred darts."[31] Plato's advocacy for the left hand was not unique.

In ancient China good fortune was attributed to the left hand.[32] "On occasions of festivity to be on the left hand is the prized position; on occasions of mourning the right hand," wrote Lao Tse, in the most widely read Chinese book in the world, the *Tao te Ching* (ca. 6th century BCE).[33] As Lydie Mepham points out, one of the difficulties in dealing with the issue of left and right in China has to do with fundamental principles in Chinese philosophy about the duality of yin and yang that have permeated Chinese thinking for over 2,000 years.*

These ambiguities were reflected in French sociologist Marcel Granet's 1933 study, "Right and Left in China," which asserted that there were exceptions to the Chinese insistence on right-handed superiority.[34] Most important, wrote Granet, *"whereas the Chinese are right-handed, the honorable side for them is the left"* (43–44). Depending on the context, the Chinese attach unequal values to the right and left hand. Rather than "absolute pre-eminence," attitudes and practices toward right and left are governed by rules of etiquette. Thus *"everything is a matter of convention, because everything is a matter of fitting"* (44).

As in China, attitudes toward left-handers in other cultures were often equivocal. As Bertrand demonstrates there has been a persistent pride among some left-handers that can be found in pottery inscriptions made by left-handers as early as the Gallo-Roman epoch. Bertrand uncovers similar evidence of "left-pride" throughout the Renaissance. He cites the Italian sculptor and architect Raffaelo da Montelupo (1505–1566) who wrote "I want to declare that I am a left-hander from birth, and that my left hand was cleverer than my right." Bertrand uncovers similar, if relatively rare, examples of left-handers who, despite dominant negative cultural values and attitudes, maintained a high self-regard about their left-handedness.[35]

Nevertheless, seemingly positive views of left-handers on closer inspection were often ambiguous. For instance, a recent examination of 50 cultures reports persistent negative connotations attached to left-handedness but finds that in the Scandinavian languages (Swedish, Norwegian, Danish, and Icelandic) the term for left (*vanster, venstre, vinstri*) translates as "pleasurable" or "helpful." Despite their seemingly positive meanings, writes the study's author, "the effort to give a good name to something considered basically bad" resembles other discriminations of minorities.[36]

In contrast with the Zulu, the Khoisans of South Africa, according to a 1938 study, seemed unconcerned with the distinction between right and left.[37] Writing in the *American Anthropologist* in 1898,

---

* Everything is yin or yang. The left hand is yin and the right one is yang. Water is yin; mountains are yang. But different sides of a mountain are yin or yang. Foods are yin or yang. So is Chinese medicine. Yang is the masculine, positive element. The sun is yang. Yin is the feminine, negative principle. The moon is yin.

another observer found some tolerance toward left-handers in Native American cultures. Left-handers seemed common, he wrote, and the terms for "left" and "left-handed" in Native American languages do not have the strongly negative meanings found in most other languages. The Mayan term for the left, *dziic*, derives from the same root as that for "soldier" or "brave" (*dziicil*).[38] Combining his examination of Native American language with tool use and painting, Daniel Brinton concluded that Native Americans were more likely than Europeans to be left-handed or ambidextrous (180). In pre-Columbian Peru, writes Michael Barsley, the Inca chief Lloque Yupanqui, whose name means "left-handed," was admired for performing good deeds.[39] Even today, among the Paiute of the American Southwest, being left-handed is viewed positively.[40]

Like the Scandinavian case above, the positive attitude toward left-handers is generally ambiguous. For instance, in the Native American Cree language, the word for left-handed is *namatinisk*, derived from *nama*, which means "no" or "not." Dutch anthropologist Albert C. Kruyt described these ambiguous attitudes and practices in his 1941 study of the Toradja of central Celebes (Indonesia). Although the Toradja refer to persons who use their left hand as "stupid," one of their divine heroes, Guma Ngkoana, was clearly left-handed.[41] Nevertheless, among the Toradja the right is associated with life and the left with death and the dead are presumed to be left-handed. Thus, the

Lloque Yupanqui. A pre-Columbian Inca chief, whose name means "left-handed," was admired for performing good deeds.

Toradja are supposed to use their left hand whenever they perform any practice related to the dead. Life remains associated with the right, and the right hand must be used for all other tasks.[42]

While right and left serve as symbols for life and death among the Toradja, this connection varies in other cultures. French anthropologist Robert Hertz's 1907 essay on death found that among the Maori death is not the binary opposite of life, nor is it connected to handedness, but rather it is a ritual that enables a transition from one state to another.[43]

It appears that attitudes and practices toward left-handers have not "evolved" from discrimination to tolerance. Rather, a more complex picture emerges in which examination of early human cultures reveals an ambivalent attitude toward left-handedness, whereby even extreme suppression of left-handers in a society can sometimes be contradicted by admiration of a particular left-hander, as in the case of the Inca chief Lloque Yupanqui or the Toradja god Guma Ngkoana. And, as in China, discrimination can follow tolerance, and tolerance may exist even where rhetoric and custom seem to stigmatize left-handedness.

ATTITUDES AND PREVALENCE

The positive or negative attitudes toward left-handers determine their reported prevalence. This is reflected in the extremely low (self-reported) prevalence of the left-handed writers among Asians answering study questionnaires compared to other groups.[44] Earlier (p. 33) we noted the extraordinary 1980s Chinese claim that less than 0.5 percent of its professional and student population was left-handed.[45] One Chinese university professor wrote recently that although he had taught thousands of Chinese students in the previous 12 years, only one was left-handed. He compared that with his experience teaching in London, where 10 percent of students were left-handed. He was certain there were no genetic differences in handedness between the two groups of students. Rather, he attributed the infrequency in China to the practice of forcing children to use their right hands, which had been abandoned in the United Kingdom, Canada, and the United States.[46]

Across the straits in Taiwan the numbers were similar. For instance, in their 1976 study of 4,143 Taiwanese elementary and college students, a Californian/Taiwanese team found that 18 percent had been frequently instructed to switch their hand use from left to right. As a result only 0.7 percent continued to write with their left hand, while 1.5 percent ate with their left hand.[47] A more focused study eight years later of adults and schoolchildren in Taiwan's capital of Taipei reported a 3.5 percent left-hand preference for schoolchildren and adults.[48] In contrast, a similar study found no significant difference in the prevalence of left-handedness between 12-year-old Chinese American elementary mission schoolchildren and University of California, Berkeley, freshmen and sophomores. While 7.6 percent of the Berkeley students threw with their left hands, so did 6.9 percent of the Chinese American children.[49] Finally, another study of Chinese American schoolchildren in Berkeley, California, found that 6.5 percent of Asian American schoolchildren used their left hands for writing.[50] While lower than the non-Asian Berkeley students, whose left-hand prevalence was 9.9 percent, the Chinese American rate was huge compared to that reported in mainland China and Taiwan.

In India the reported rates of left-handedness are higher than those in China but significantly lower than rates among European, Canadian, and American children of the same age.[51] A 2001 follow-up study of 10 unimanual activities by New Delhi (India) private school students, ages 4 to 11, found a left-hand prevalence of 3.2 percent.[52] A second study of 6- to 18-year-old Indian students at another private school, north of New Delhi, reported a 4.25 percent left-handed prevalence.[53]

"Cultural and environmental factors," write French neuropsychologist Maria De Agostini and her colleagues, "change 'natural' hand preference in three ways." First, they might only affect one activity, such as eating or writing, but not influence other common unimanual tasks.[54] Second, those who experience temporary injuries to their dominant hands tend to become "ambi-handed" rather than totally shifting to their noninjured hand.[55] Finally, attitudes toward left-handedness can directly influence left-handed prevalence in any society.[56]

The first and third of these cultural influences have shaped Chinese practices toward handedness, but these forces are more dynamic than usually portrayed. Not least of all is the social and economic cost of making adjustments for 10 percent of the population. As the *China Daily* recently reported, "In China kids are all taught to write with their right hands. If they pick up a pencil with their left hand, the teacher will put it in their right. It's really just a matter of practicality. In the US, you have left-handed desks, left-handed guitars, and all sorts of other left-handed devices, but in China we have none of the sort."[57]

However, a number of reports agree with the observation that there are many more left-handers in Shanghai than in Guangdong. Thus, the report concluded that forced switching is declining in China's most modern cities.[58]

"Of course," writes China expert Mepham, there are "left handers in China ... but maybe not in the past. In fact being a left hander nowadays is a good thing when playing ping pong, tennis etc. [as] it must be true in our western sphere also today." But, Mepham concedes, forcing left-handers to write with their right hands persists. "A Chinese friend," writes Mepham, "tells me that all the left-handers he knows, actually can also write ... with their right hand!"[59]

Geneticist Krishna R. Dronamraju uncovered a higher than expected left-handed prevalence among Andhra Pradesh (Indian) tribals. This persuaded Dronamraju that pressures to switch left-handers to right-handedness were more recent in India than generally supposed. Dronamraju urged more investigations of the rate of left-handedness in other tribal and aboriginal cultures.[60]

Researchers from Hertz to Zverev have insisted that cultural values significantly influence the epidemiology of hand preferences.[61] According to psychologist Clare Porac and colleagues, prevalence differences between and among societies worldwide are more strongly correlated with cultural than biological pressures. Nevertheless, they find that variability in handedness *within* cultures reflects neurological and genetic forces.[62]

Technological innovations, from early human tools to contemporary scissors, can openers, and student desks have created complications for natural left-handers and resulted in reduced left-handed prevalence.[63] The spread of literacy, especially of writing, initially also increased pressures on left-handers to become right-handed writers.[64] It probably played a role in the decline of prevalence of left-handers in the United States and Britain in the nineteenth century, described by a number of studies and discussed in depth in chapter 3.[65] One factor that recently has encouraged an increased prevalence of left-handedness is the worldwide sports craze, in which left-handed athletes have an advantage in a number of sports, including baseball, tennis, football, and soccer; and, in China, left-handed ping-pong players have increased the tolerance for left-handers.

Although there has been an ambiguous view of left-handers throughout human history, negative have mainly trumped positive attitudes. This supports Hertz's conclusion that the preeminence of the right hand and subordination of the left have been determined by culture rather than by biology.[66] In large part, this is because, as Hertz argued, humans have and continue to seek explanations for the unknown in a binary framework of the sacred versus the profane. Like their biological asymmetries, humans created cultural asymmetries, which assigned different tasks to different social categories. Among the most powerful cultural asymmetries were sex, gender, color, and right- and left-handedness. The latter specifies which tasks are assigned to right and left hands. Some of these, like eating, are associated with the sacred and restricted to the right hand. Others, like cleaning ones' feces, are assigned to the left hand, which symbolized the negative, the profane. This binary opposition of left and right is so entrenched in our language and social customs that even when specific beliefs that restrict left-handers are neutralized, as they increasingly seem to be, hidden antipathies have emerged clothed in educational, scientific, and medical theory justifying the retraining of left-handers.[67] One need look no further than American politics to confirm the persistence of characterizing difference as a contest between good and evil.

There may be increased tolerance, but for approximately two-thirds of the world's population, being born left-handed exposes a person to discrimination and often to retraining. Although its advocates often insisted that retraining not only was harmless but also improved a child's educational performance, its opponents came to the opposite conclusion. Much like the testimonies of the retrained that opened this chapter, their supporters identified lifelong harms to the forcibly switched, not least of all stuttering. This is the focus of chapter 5.

5

CHANGING HANDS,
TYING TONGUES

Herein lies the crime against the left handed: When a child at home or at school shows a decided tendency to use the left hand in preference to the right, persistent effort is generally made to force such child to use the right hand. . . . Many cases of stammering and stuttering in children are, no doubt, the result of the forcing of constitutional left-handed individuals to become right-handed.

HARVEY ERNEST JORDAN (1922)

Sinistrality is thus nothing more than an expression of infantile negativism and falls into the same category of other well-known reactions of a similar nature, such as contrariety in feeding and elimination, retardation in speech, and general perverseness. . . . Children should be encouraged in their early years to adopt dextrality. . . . The alleged dangers of retraining are nonexistent.

ABRAM BLAU (1946)

M any children in my elementary school in the 1950s stuttered. They were ridiculed by other pupils and, sometimes, teachers. Because their tortuous responses to questions disrupted the flow of classroom discussions, teachers rarely called on them. Embarrassed and humiliated, children who stuttered sought to remain inconspicuous.[1] A colleague and contemporary of mine recently confessed that as a stutterer he had spent his childhood avoiding humiliations, not always successfully. Although he knew for the past several years that I was researching left-handedness, only after watching the film *The King's Speech* did he reveal that he had been a left-hander switched to writing with his right hand. The film, which he had viewed multiple times, provided him with the courage to disclose his past mortifications. Today, there are many fewer child stutterers in American classrooms and fewer attempts to force hand switching in public schools. In much of the non-Western world, however, where switching remains a common practice, stutterers are as common as they were when I was a child. Does hand switching increase the risk of stuttering? Or were left-handers more prone to language disorders and thus to stuttering than the right-handed population?

OPPOSING VIEWS

Although they disagreed about almost everything connected with handedness, criminologist Cesare Lombroso and anthropologist Robert Hertz both believed that no attempt should be made to transform left-handers into right-handers—for different reasons. Lombroso was convinced that left-handedness was an "atavistic" or "primitive" trait that could not be altered.[2] Therefore, it was unnecessary to change the hand preference of left-handers. In contrast, rather than anticipating the eventual extinction of left-handers, Hertz urged encouraging left-handedness and left-hand use among both right- and left-handers. The engagement of both brain hemispheres would result in ambidextrous individuals, more intelligent and creative than the unimanual population, he averred.[3]

In traditional cultures restriction of the left hand was based on deep-seated beliefs that its use was profane. By the twentieth century similar discriminations were justified in the West by scientific and medical discourses. Most twentieth-century British and North American educators and psychologists rejected Hertz's argument, siding instead with Lombroso's portrayal of left-handers as defective. But contrary to Lombroso, who believed that left-handers could not be "cured," these experts advocated forcing left-handers to become right-handed. Using a "scientific" rationale, they insisted that children should be compelled to write with their right hands. Rejecting the notion of natural or inherited left-handedness, these experts asserted that a child's decision to rely on his or her left hand was a reflection of a stubborn and defiant personality that could best be corrected by forcible switching.

In the 1920s, a number of British and American pediatricians insisted that left-handers displayed some degree of retardation and feeblemindedness. Left-handed children, wrote the Scottish physician William Elder, were severely handicapped; they always were below average in intelligence and often feebleminded.[4] Elder and others urged early switching of left-handers, to reduce the potential harms of left-handedness. Oddly, Elder also believed that "ambidextrous children were above the average in mental capacity."[5]

In contrast, other experts tied learning disabilities among left-handers to retraining. These educators and physicians who disapproved of forced switching saw left-handedness as inherited and natural, and they warned of the negative consequences of this practice, especially stuttering. Probably the best-known case of stuttering attributed to hand switching is that of Duke of York, the future King George VI, who had been forced to write with his right hand. The duke's struggles to overcome stuttering were brought anew to public awareness in the 2010 film *The King's Speech*, based on the book of the same name.[6] In his 1958 biography, Sir John Wheeler-Bennett wrote that the future king "did not stammer when he first began to talk"; his stuttering began "during his 7th and 8th years. It has been attributed to his being naturally left-handed, and being compelled to write with his right hand. This would create a condition known in psychology as a 'misplaced sinister' and may well have affected the speech."[7] It is impossible to be certain that hand-switching was at the root of the king's stuttering, but circumstantial evidence provided by similar cases was used by those who questioned the wisdom of "retraining" natural left-handers to become right-handers.

The Duke of York (the future King George VI) playing tennis left-handed, 1921. The duke was forced to use his right hand for eating and writing and soon after, at age 7, he began to stutter.

PROTESTS AGAINST SWITCHING

As early as the 1920s, some observers began to suspect that the practice of retraining left-handers negatively affected their speech develop-ment.[8] A 1917 survey of the schools for delinquent and feebleminded by Clark University educator Laura G. Smith uncovered a high per-centage of left-handers.[9] Although she conceded that her finding on delinquency and learning disabilities appeared to support the tra-ditional negative views of left-handers, Smith also found that many left-handers were extraordinarily talented. Forced switching, she lamented, not only robbed the world of potentially talented individ-uals but also ironically resulted in the classification of resistant left-handers as mentally deficient.[10]

Others joined Smith in protesting retraining. Writing to parents and teachers in 1917, physician J. J. Terrell characterized forced retrain-ing of left-handers as a "ruthless" practice imperiling a child's men-tal development.[11] Harvey Ernest Jordan, professor of histology and embryology at the University of Virginia, specifically rejected Lombro-so's claims that left-handers were criminals and defectives. Rather, he deplored *forced hand switching* as criminal, because it placed a child at risk of developing serious speech defects, especially stuttering.[12] At first it may seem surprising that Jordan, an enthusiastic eugen-icist who was convinced that racially "superior" humans should be encouraged to reproduce while "inferiors" should be discouraged or even subjected to sterilization, should disagree with Lombroso's views about left-handers. However, as we will discuss more fully in chapter 6, Jordan saw left-handers as a natural result of Mendelian inheritance. Switching them to be right-handers, Jordan believed, violated genetic laws, resulting in pathology for those subjected to retraining.[13] Statistical investigations confirmed Jordan's warnings of a much higher incidence of stuttering among retrained left-handers.[14]

The most comprehensive of these were three studies produced by J. P. Ballard in 1911–12. Ballard's findings led him to conclude that nothing could be gained by trying to make a left-handed child right-handed. "On the contrary," he wrote, "there is a grave risk of stammer-ing resulting from the change." Ballard's findings led him to conclude that nothing could be gained by trying to make a left-handed child right-handed.[15] Published in an obscure pedagogical journal, Ballard's studies would have gone unnoticed if not for their republication in the influential 1914 textbook *The Hygiene of the School Child*, by Stanford University psychologist Lewis M. Terman. Based on his interpretation of Ballard's data, Terman concluded that forcing left-handers to become right-handed increased the risk of stuttering by a factor of four. This evidence, Terman told his readers, demonstrated the risk associated with retraining. Left-handers, he concluded, should be left alone.[16]

Although Terman's views were influential, many remained skeptical. For instance, Elder argued that it was "more reasonable to suppose that the stammering which may have come on during the training of a child to use the right hand is not the result of such training, but rather the derangement often associated with left-handedness."[17]

Elder's views and views like his, while persistent among practitioners, contrasted with the findings of researchers who believed that teaching the left-handed child to use his right hand resulted in speech disturbances.[18] In 1932 Ira S. Wile, president of the American Orthopsychiatric Association and former New York City commissioner of education, reviewed the evidence about the connection between stuttering and other language disorders to forced hand switching and concluded that behavioral disorders were more common in retrained left-handers than those left alone.[19] In his book *Handedness: Right and Left*, published two years later, Wile claimed that even temporary hand switching increased the risk of speech impairment, including mild stuttering. Among those natural left-handers who were "successfully" switched, the results were more debilitating, including severe stuttering, problems with arithmetic, and difficulty in reading.[20]

In his 1932 article Wile appended case histories of 20 children who had been switched from left-handers to right-handers. The group presented an array of physical and behavioral difficulties ranging from bed-wetting, stuttering, and learning difficulties to disruptive behaviors such as lying and stealing. When the children were switched back to their natural hands, their negative behaviors, stuttering, and bed-wetting were eliminated.[21] One 9-year-old boy, for example, was "uncontrollable, inattentive at school, untruthful, stealing, and retarded." Taunted by his classmates, the boy wrote poorly with his right hand and had difficulty reading. After he was allowed to return to his natural left-handedness, the boy's attention and self-control improved, his stealing ceased, and he was promoted (case 3, 52). Forced to repeat three grades, another 9-year-old boy, characterized as stupid,

Ira S. Wile (1877–1943), the president of the American Orthopsychiatric Association and former New York City commissioner of education, concluded that forced hand switching resulted in stuttering and other behavioral disorders.

had difficulty with reading and social interactions. He began stuttering when he entered school, where he was forced to convert to right-handedness. When he was reconverted to left-handedness, his stuttering diminished and his reading skills improved remarkably (case 12, 53). A 10-year-old with an IQ of 135 who had been characterized as lazy and with poor English skills was also reconverted, whereupon his behaviors and attitudes were transformed at school and at home (case 7, 52). Similarly, a 7-year-old boy, when allowed to resume writing with his left hand, "increased learning power" and completed his assignments much more quickly (case 8, 52).

Recognizing that pediatric developmental disorders, including stuttering, resulted from multiple causes, Wile nevertheless concluded that "if left-handedness were more persistent, there would be fewer victims of violent speech disorders." Retraining, he insisted, not left-handedness, was the cause of these disorders.[22]

Wile's observations that returning switched left-handers ameliorated stuttering and other learning disabilities were given authoritative credence by two influential University of Iowa researchers, psychiatrist Samuel Torrey Orton (1879–1948) and psychologist Lee Edward Travis (1896–1987). Orton was founding director of the State Psychopathic Hospital in Iowa City and chairman of the Department of Psychiatry at the University of Iowa, College of Medicine, while Travis, a founder of the profession of speech pathology, was director of the university's speech clinic. From the late 1920s until the 1950s, both collaboratively and separately, Orton and Travis published on the connection between stuttering and forcing natural left-handers to write and perform other tasks with their right hand, and between stuttering and those who relied on both hands for these tasks, often categorized as being ambidextrous.[23]

While Travis focused on speech pathology, Orton connected their findings on stuttering to the cause of reading disorders. Orton was persuaded that stuttering resulted from weak laterality.[24] That is, in most humans, one cerebral hemisphere, usually the left, was dominant for language and motor function. In a subset, especially among those with speech and reading disorders, Orton hypothesized, the hemispheres were symmetrical with neither side being dominant. Orton believed that cerebral confusion was implicated in reading disabilities, which he gave the label "dyslexia."[25] For the vast majority of children, said Orton, heredity determined which hemisphere was dominant for language, with most children being left-brained (thus right-handed) for language and writing.[26] "One side of the brain," wrote Orton, "is all important in the language process and the other side either useless or unused."[27] Orton reported that a number of children who had difficulty recognizing letters either had been switched from being left-handed or had not evidenced dominance by either hand.[28]

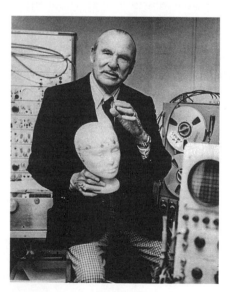

Lee Edward Travis (1896–1987), a founder of the profession of speech pathology, was director of the University of Iowa's speech clinic from the late 1920s until the 1950s. Travis collaborated with his colleague, psychiatrist Samuel Torrey Orton (1879–1948), on the connection between stuttering and forcing natural left-handers to write and perform other tasks with their right hand.

Travis emphasized that there were multiple causes of stuttering but, nevertheless, he found that approximately 50 percent of stutterers were left-handers retrained to be right-handers. Many of the stutterers appeared to be ambidextrous, but according to Travis, even much of the ambidexterity was attributable to parents' and teachers' efforts to convert left-handed children to right-handedness.[29] Travis noted that the same, albeit rarer, scenario sometimes occurred in reverse to natural right-handers. His approach to curing stutterers was to enable one hemisphere to be in complete control. So-called ambidextrous children were at greater risk of becoming stutterers than strongly lateralized children. Because most children began stuttering before the age of 8, Travis urged early interventions. For a majority of stutterers the way to do this was by reestablishing their left-handed dominance (186).

In 1937 King George's assistant private secretary, Sir Alan Lascelles, asked the king's speech therapist, Lionel Logue, whether hand-switching might cause speech impediments. Logue responded that stammering might be reversed if the person was switched back to his or her left hand; but, he said, such an intervention could not help the king because, at 42, he was too old.[30] Apparently, Logue was not familiar with the work of researchers at the University of Iowa, discussed below, who had published numerous case studies claiming that switching back to left-hand use could cure or reduce stuttering even among those much older than the king.[31]

As early as 1931 Travis published case studies of his patients illustrating that restoring left-handedness was a successful treatment for stuttering. There was the 18-year-old left-handed male whose stuttering began after he was forced to write with his right hand. Although the patient had a family history of left-handers and stutterers, Travis rejected claims that left-handedness was more common among people with physical and mental defects, pointing to the patient's recovery from stuttering six months after being restored to left-handedness.[32]

This intervention was applied in Travis's clinic and elsewhere, with good results. One member of the Travis staff, Dr. Leo Fagan, reported that 33 putatively right-handed stutterers were required to use the left hand for daily activities. After two years, 9 of the 26 patients spoke completely normally, while 15 others had improved substantially.[33]

In 1935 two of Travis's former graduate students, Wendell Johnson and Lucile Duke, published an elaborate series of case reports of 16 stutterers, 10 males and 6 females, ages 5 to 71, 14 of whom were left-handers made to write with their right hands.* The 6 females ranged from 10 to 30 years in age. All 16, due either to accident or to design, reverted to writing with their originally dominant hand. Subsequently, all the stutterers were either cured or their stuttering was substantially ameliorated. Johnson, a stutterer himself, went on to have a long and distinguished career as a psychologist and speech pathologist at the University of Iowa. Although Johnson had no evidence that he had been forced to become a right-hander, he attempted to cure his own stuttering by switching to left-handedness. His attempt failed.[34]

The first case was H. T., a 37-year-old man who, from infancy, had been trained by his mother to use his right hand, despite his preference for his left hand. According to his mother, H. T. barely spoke until he was 4½ years old, a situation that his mother attributed to her son's obstinacy. When H. T. entered school at age 6, his teachers reinforced the right-hand training. Although H. T. was talking he had difficulty with pronunciation. Forced to write with his right hand, H. T. began stuttering, which became increasingly severe during adolescence. H. T.'s parents and teachers considered him completely right-handed, even though he used both hands equally in untrained activities and preferred his left hand for most manual activities. H. T.'s stuttering continued until, at age 32, he suffered a severe injury when his right hand was caught in farm machinery. H. T. was forced to rely on his left hand for all activities. Over the next two years H. T.'s speech improved and by the third, his stuttering disappeared.[35]

Johnson and Duke ruled out suggestion as the cause of H. T.'s recovery from stuttering. "It is of special interest," they wrote, "that the patient professed essential ignorance as to the reason for the disappearance of the stuttering. When it was suggested that perhaps the change of handedness was responsible, he expressed astonishment and was loath to accept the explanation" (114–15).

R. S. was a 10-year-old natural right-handed boy whose speech had developed normally. At age 6 he contracted poliomyelitis and was forced to use his left hand for the next five months, during which he

---

* Six of the cases were aged 5 to 10, three were 12 to 18, and six were 20 to 71 (20, 26, 30, 37, 39, and 71).

developed severe stuttering. As R. S.'s right hand began to improve, his father tied the boy's left hand, so he was forced to use his right. After three months the stutter diminished: the more R. S. relied on his right hand, the more fluent his speech became. The stutter soon disappeared (116).

And so it went through the remaining 14 patients. Each patient's stuttering began after forced retraining. With the exception of R. S. and a 71-year-old man (case 9) who stuttered after a stroke robbed him of the use of his right hand, all were left-handers retrained to use their right hand for writing. Typically, the patients were like C. G., a 10-year-old who had been writing with his left hand when at age 7 his teacher forced him to use his right hand. Shortly after, he began to stutter. Brought to the Iowa Speech Clinic in April 1931, C. G. was instructed to switch back to his left hand for all activities, which he did—and thereafter his mother reported that the boy was talking normally, reading better, and writing better, and was more sociable and happier since he had reverted to using his left hand (117–18). Based on these case reports, Johnson and Duke concluded that returning a patient to the preferred hand caused stuttering to disappear.[36] Thus, Johnson was convinced that stuttering and other fluency disorders were learned rather than a result of a physiological defect.[37]

Two years later, in his widely used textbook *Reading, Writing, and Speech Problems in Children*, Orton wrote that reversal of training was often followed quickly—within a matter of weeks—by the disappearance of the stuttering.[38] But Orton also warned his readers that these successes had given rise to the misleading belief that all stutterers were intended to be left-handers and ought to be trained to predominantly use their left hand. Although many stutterers demonstrated no clear preference, some, especially those with a very strong heredity of stuttering, were exclusively right-sided from birth (195).

These views were endorsed and expanded by Travis's former student Bryng Bryngelson (1892–1979), the director of the speech clinic at the University of Minnesota, who also served as president of the American Speech-Language-Hearing Association (ASHA) from 1943 to 1944. Bryngelson identified two types of stutterers. The first were well adjusted and seemed to be untroubled by their stuttering.[39] The second group were deeply distressed by their stuttering. These individuals, Bryngelson noted, were weakly lateralized and, as a result, had great difficulty integrating the bilateral elements of speech (195).

Focusing on the second group Bryngelson reported that 81 percent of his 127 stuttering patients had been switched from left- to right-handers. According to Bryngelson, their stuttering could be traced to the ambidexterity of some of them and to the fact that many strong left-handers resisted retraining and had only incompletely adopted right-handedness.[40] Further studies confirmed these earlier results.[41] From these clinical studies Bryngelson concluded that "the central mechanism of a stutterer appears to be in a state of ambilaterality, neither the right nor the left cerebral hemisphere exercises a dominant lead control over the other." Bryngelson was not dogmatic about his findings, however. He admitted that "there is no reason to believe that, because the search for the etiology of stuttering has obtained for nearly 2,000 years this is the age in which speech pathologists will find its ultimate and final truth."[42]

Other researchers in the 1940s reported similar findings. Reviewing the past two decades of research on "various branches of science to the study of handedness and its relation to speech," renowned London neurologist W. Russell Brain (1885–1966) reported in the *Lancet* in 1945 that "there is considerable evidence correlating stuttering with anomalies of handedness."[43]

SKEPTICS

Not everyone was convinced. For them, both practical and developmental reasons necessitated switching left-handers. For one thing, as Beaufort Sims Parson observed in 1924, the classroom is designed for right-handed pupils, and this puts left-handers at a disadvantage: the lighting is for right-handers and creates eye strain in left-handers; left-handed students bump into their right-handed classmates and contort themselves to climb into their seats. Parson concluded that persistent left-handers and the opponents of switching them posed a serious obstacle for primary schoolteachers.[44] Parson rejected the connection between stuttering and retraining of left-handers except among those switched at an older age whose cerebral processes have by then become fixed (27). He pointed to a four-year intensive campaign by the Elizabeth, New Jersey, public schools (1918–1922), in which 184 of 250 left-handed writers were retrained, with no subsequent negative effects (100).[45]

Pointing to the same study 14 years later, University of North Carolina educator K. C. Garrison wrote that the Elizabeth schools did not have a single case of defective speech resulting from hand switching.[46] For Garrison the culprits were childhood neuroses coupled with unsophisticated retraining methods; together, they accounted for the few cases of subsequent speech pathology (328–29). The problem, Garrison wrote, was that left-handers "were more psychopathic than the dominantly right-handed" (331). Although Garrison noted persistent reports of low scholastic performance among left-handers, he cited a 1936 study of 200 elementary schoolchildren that "showed a lack of relationship between handedness, eyedness, or mixed hand-eye dominance and the degree of reading efficiency" (330).

Psychoanalytic Responses

Garrison's rejection of a connection between forced switching and stuttering reflected the growing dominance of psychoanalytic psychiatry in the post–World War II era. There were, of course, numerous psychiatrists in the 1940s and 1950s who continued to believe that stuttering was an organic disorder. Academic psychiatry, however, especially as taught to psychiatry and neurology residents, was dominated by psychoanalytic theory.[47] This translated into control of the editorial boards of professional psychiatric journals. It was impossible for psychiatrists critical of psychoanalytic theory to publish in American psychiatry journals.[48]

For psychoanalysts stuttering resulted from unconscious early childhood conflicts.[49] From an orthodox Freudian perspective the left-handedness of stutterers was evidence of a combination of poor parenting and childhood rebellion. Thus Boston psychoanalyst Isador H. Coriat (1875–1943) explained in 1931 that stuttering was caused by "the fear of the ego being overwhelmed by the all-powerful autoeroticism." Stutterers, wrote Coriat, were desperately reenacting their nursing fantasies by a repetition-compulsion of speech involving their tongue, lip, and jaw movements. From this perspective stuttering was a psychoneurosis similar to narcissism. Given its causes, stuttering could only be treated by experienced psychoanalysts. Other interventions, he warned, such as those practiced by Orton, Travis, and their followers, should be discouraged.[50]

By the 1940s, the growing influence of European émigré psychoanalysts combined with newly trained American psychoanalytic psychiatrists to reframe the understanding of a wide variety of pediatric
developmental and learning disabilities, including autism, hyperkine-
sia, dyslexia, Tourette syndrome, and stuttering.[51] Like Coriat earlier,
these psychoanalysts rejected the widely held view that stuttering resulted from neurophysiological disturbances. Instead, they emphasized
unconscious mechanisms that, when unearthed, revealed repressed
"homosexuality, masturbation, anal eroticism, identification with the
father, return to baby speech, humiliation, secretiveness, etc."[52]

Expanding Coriat's emphasis on oral fixation, psychoanalyst Else
Heilpern also emphasized the role of anal fantasies as a cause of stuttering. Heilpern endorsed and translated into English the views of
Viennese (later Berlin) psychoanalyst Otto Fenichel,[53] that stuttering
was "a pregenital conversion neurosis presupposing an erotization
of the speech function; . . . mostly anal, and underlying oral character; its aims are almost constantly of an exhibitionistic and sadistic
nature."[54] Fenichel had settled in Los Angeles in 1938. His views on
stuttering were imported into his authoritative 1945 English-language
psychoanalytic textbook, *The Psychoanalytic Theory of Neurosis.*[55]

This psychoanalytic orthodoxy penetrated every aspect of educational psychology. The disappearance from the literature of the connection between stuttering and forced hand switching was one of
the collateral results. A turn toward psychoanalysis is evident in the
arguments of New York City school chief psychiatrist Abram Blau's
widely influential 1946 text, *The Master Hand: A Study of the Origin
and Meaning of Left and Right Sidedness and Its Relation to Personality
and Language.*[56] Blau cautioned that the increasing acceptance of
left-handedness by parents and teachers would result in severe developmental and learning disabilities (67–71). To avoid such outcomes
he insisted that as soon as their left-handed preference was noticed,
children should be retrained to use their right hands (184–85).[57]

Blau identified three general causes of left-handedness. The first,
developmental deficiency, included children with physical and learning disabilities (87). The second, defective educational theories, was
often exacerbated by left-handed parents (82, 89). Finally, there were
the neurotic and oppositional children themselves, for whom left-
handedness was "a symptom or manifestation of an attitude of opposition or negativism along with such other signs as disobedience,

refusal to eat, temper tantrums, rebelliousness, etc." (91). Left-handed-ness was grounded in psychoanalytic concepts concerning the conflict between infantile drives and the educational and oppositional role of the child's mother. Inconsistent parenting made the child's resistance worse and resulted in a personality disorder marked by pathologi-cal negativism that manifested itself by the child's refusal to adopt right-handedness. This resistance, wrote Blau, arises from infantile emotional immaturity and "falls into the same category of other well-known reactions of a similar nature, such as contrariety in feeding and elimination, retardation in speech, and general perverseness" (112–13). The increase in left-handers, wrote Blau, resulted from the indefensible practices of dissident parents and educators who failed to appreciate the dangers to the mental health that accompanied per-sistent left-handedness (92).

Given his perspective, Blau particularly deplored the practice of experts advising parents to allow the child to choose the side for himself or herself. This allegedly progressive practice, Blau warned, was harmful to the child, as it undercut the cultural value of right-handedness (90). Only education and behavior modification could avoid left-handedness and its disabling sequelae (5, 19, 21, 93, 184). Blau's psychoanalytic views were widely endorsed by educators in the 1950s; they cited and reinforced Lombroso's earlier claims that left-handedness was abnormal and that those who resisted retraining were at great risk of learning disabilities.

Indeed, Blau's views were emblematic of psychoanalytic psychi-atrists in the 1950s and 1960s. Although I had assumed that the psy-choanalytic explanations for stuttering were no longer taken seriously, even by psychoanalysts, I was astonished to hear a member of a family psychiatry working group to which I belonged describe, in 2012, a stuttering patient as driven by a narcissistic need to make himself the center of attention, despite the fact that stutterers regularly reported that their main desire was to avoid any notice whatsoever. Of course, as I had learned in my psychoanalytic seminars, the denial of a desire actually exposes its unconscious opposite. In this case, the stuttering putatively betrayed an unconscious wish to be noticed, since it forced others to focus attention on his behavior.

Mainly these theories furthered claims that left-handedness was abnormal and that those who resisted retraining were at great risk of learning disabilities. Thus, Brooklyn College educator and psychologist

Gertrude Hildreth (1898–1984) insisted that left-handed children should be converted to right-handers because the learning environment was structured for right-handers. Failure to do so would have negative consequences.[58] Similar arguments made their way into educational journals. In a 1952 article in the *Journal of Educational Research*, educator Kenneth Martin urged his readers to discourage the use of the left hand in all childhood activities from shaking hands to the use of "rattles, spoons, pencils, play things, utensils," and so on. But Martin conceded that not all children could be made into right-handers.[59] Moving away from totalistic positions of orthodox psychoanalysis, Martin proposed a more flexible response, including classroom furniture, tools, and lighting that would allow left-handers to operate at the greatest possible efficiency. Martin (531) cited a 1943 *Parent's Magazine* article called "It's No Fun to Be a Southpaw."[60]

TOWARD A MIDDLE GROUND

By the late 1950s, while educational psychologists had generally rejected Blau's draconian attitude toward left-handers, most were skeptical that there was a direct cause and effect between retraining left-handers and the onset of stuttering. In her authoritative 1957 investigation of educational implications of left-handedness in Scotland, educational psychologist Margaret M. Clark concluded that "stuttering may result from changed handedness, but whether it does or does not will depend to a great extent on the procedure adopted in effecting the change."[61] Stuttering was not so much a result of the attempted changes, wrote Clark, but of the "emotional disturbance" which might arise from the child's resistance to being switched in the first place (51). What was critical was how old the child was when the change in handedness was initiated. It was desirable to switch left-handers early, because older children's brains were more highly developed and would experience greater stress, which itself increased the risk of stuttering (51). Based on her review of the literature and a comprehensive study of 330 Scottish elementary pupils, Clark was persuaded that in the vast majority of children "no essential difference was apparent between left-handers and right-handers except for their use of a different hand" (201). She urged educators and parents to identify retrained left-handers and return these children to their dominant hand. "Few people realize just how common it still is for left-handers to be encouraged or forced to use their right hand," wrote Clark in 1957. "Some left-handers are permitted to write with the left hand, but by no means all" (38).

Brooklyn College speech pathologist Oliver Bloodstein, who obtained his PhD in 1948 at the University of Iowa with a dissertation directed by Edward Travis, was skeptical of the connection between stuttering and forced hand switching. He nevertheless conceded that it was "difficult to read certain case reports of children who began to stutter after a shift of handedness, or to study accounts of certain adults who experienced fluency disturbances after enforced use of the nonpreferred hand, without being impressed by the possibility that laterality is a factor in some cases" of stuttering.[62] Supporting that possibility was a 1966 report by Philadelphia neurosurgeon R. K. Jones of four stutterers who developed aphasias. Stuttering ceased in all four after surgery in one hemisphere, persuading Jones that Travis was correct to connect stuttering to weak asymmetry.[63]

Reviewing the literature of the previous half century, in 1983 New Zealand psychologists Michael Corballis and Ivan L. Beale found "that stutterers often show anomalies of cerebral lateralization," including reversal of cerebral dominance, which could have resulted from forced switching of left-handers. Citing the patients described by Jones in 1966, Corballis and Beale noted that some stutterers manifested "bilateral control over speech, leading to interhemispheric conflict."[64]

In contrast with Corballis and Beale, psychologist and handedness historian Lauren Harris described the evidence for this connection as weak. Yet, Harris admitted that there had been no new investigations of the stuttering/hand switching hypothesis that refuted the Iowa findings of the 1930s and 1940s. That was because, wrote Harris, subsequent studies had focused on the wider issue of whether left-handers had greater risks for learning disabilities than right-handers. Thus, Harris conceded that it was possible that forced hand switching was a causal factor for stuttering, especially among some left-handers.[65]

THE DECLINE OF STUTTERING?

Perhaps the primary reason that hand retraining is no longer a subject for investigations into the etiology of stuttering is that the practice of hand switching has significantly diminished, especially in North America and Britain, the sites for most dysfluency or stuttering research. By the late 1950s North American and British educational psychologists had generally rejected forced switching of left-handers. According to Australian psychologist C. J. Brackenridge, there was an increase in the number of left-handed writers in English-speaking countries from the late nineteenth to the mid-twentieth centuries. Brackenridge finds the most dramatic increases first among high socioeconomic-status children. For instance, the reported rate of Australian or New Zealand left-handers increased from 2 to 13 percent between 1880 and 1969. A similar pattern emerged slightly later in the United States.[66] Corballis attributed this increase to a growing acceptance of left-handers, which is reflected in the increase of stores that sell tools and utensils for left-handers.[67]

Conclusion

Although it is difficult to obtain reliable data on the incidence and prevalence of stuttering even in Britain and North America, where the practice of retraining left-handed children has almost disappeared, there is evidence that suggests that the prevalence of stuttering has likewise declined.[68]

Even if hand switching accounted for a significant percentage of stuttering, it could not account for the etiology of all stuttering; indeed, none of the Iowa researchers in the 1930s and 1940s made such a claim. Rather, Orton, Travis, and their colleagues insisted that an understanding of the likely mechanisms that underlay the connection between retraining and stuttering opened the door to the wider issue of the role of weak laterality and stuttering.

Orton had concluded that the reduced cerebral dominance that Travis had seen as the cause of stuttering also was responsible for the reading and writing deficits found in dyslexia. Reexamining these claims in 1991 Corballis reported that, while the statistical data failed to corroborate Orton's hypothesis, patient reports and memoirs provided compelling evidence in its favor. "Sometimes," wrote Corballis, "the dead hand of statistics is no more to be trusted than the embrace of the overenthusiastic theorist."[69] Based on his review of patient cases and memoirs, Corballis was persuaded that dyslexia was connected to ambidexterity, anomalous handedness, and left-right confusion (201). In this context Corballis reexamined Travis's claim that forced hand switching of left-handers caused stuttering. Distinguishing the weak motor (*praxia*) laterality found in dyslexics with the linguistic confusions evident in stutterers, Corballis concluded that "it would be more pertinent to examine the relation of stuttering to weak cerebral dominance rather than to handedness" (204).

Although current stuttering experts remain skeptical about the Iowa group's claims, following Corballis's suggestion, a number of recent investigations using new imaging technologies have reopened the possibility. While their conclusions are tentative, they are intriguing. For instance, in an MRI study of 32 adults, neurologist Anne Leigh Foundas and colleagues found that an "anomalous anatomy" in the speech-language areas of the brain, known as the perisylvian area, increased the risk for stuttering.[70] Foundas and her colleagues have continued these investigations and report that adults with persistent developmental stuttering display atypical cerebral laterality.[71] Other investigators have reported similar findings. Reviewing a number of these reports, the German team of Christian Büchel and Martin Sommer concluded that stuttering is connected to disturbances of signal transmission between the hemispheres.[72] This signaling disturbance, write Büchel and Sommer, "also explains why the normal temporal pattern of activation between premotor and motor cortex is disturbed and why, as a consequence, the right hemisphere language areas try to compensate for this deficit" (162). More recently, a Swedish team using transcranial magnetic stimulation (TMS) on 30 subjects, combined with findings of other recent imaging studies, found a connection between stuttering and left-hemisphere motor impairment.[73] Like so many before them, they concluded that weak laterality of speech is the most likely mechanism in stuttering (5).

These findings mesh with similar claims by Orton, Travis, Bryngelson, and their students in the 1930s. Thus, current investigators have found that those stutterers who are able to rely on their right hemisphere for language can become fluent.[74] And, like Orton and Travis in the 1930s, twenty-first-century investigators have concluded that genes play a role in the risk of stuttering. PDS (persistent developmental stuttering) requires the interaction of environmental factors as well (163). Of course, all of these recent imaging studies are suggestive rather than conclusive. As a team of Swiss researchers pointed out,

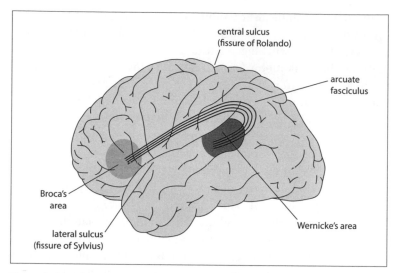

central sulcus
(fissure of Rolando)

arcuate
fasciculus

Broca's
area

lateral sulcus
(fissure of Sylvius)

Wernicke's area

The perisylvian cortex, associated with language and speech, is located around the Sylvian fissure.

while recent studies are strongly suggestive that PDS is connected to anatomic anomalies, it is not limited to speech and languages (perisylvian). Rather, it included prefrontal and sensorimotor areas as well. Nevertheless, the question remains whether these anomalies are the cause or consequence of stuttering.[75]

For Orton, Travis, Bryngelson, and their students, weak lateralization was a cause rather than the result of stuttering. Often they found that reversing forced hand switching cured or relieved stuttering. In the West, forced training of left-handers to switch to their right hands has declined, as has the incidence of stuttering. Is this a consequence of the disappearance of training, or is it a coincidence? The best way to answer that question would be a carefully constructed randomized clinical investigation to determine once and for all whether or not the intriguing hypotheses, developed and employed in the University of Iowa clinics in the 1930s and 1940s, have any validity. This would not simply serve an antiquarian interest, because, outside the West, most societies and cultures continue to practice forced retraining of left-handers. As the British and North American experiences recounted in these pages suggest, such practices are not only discriminatory, they also may impair the physical and emotional development of left-handers, who constitute 10 percent or more of the world's population.

Reviewing the studies supporting and critical of the connection between the onset of stuttering and forced switching of left-handers in his 1971 authoritative textbook *The Nature of Stuttering*, speech pathologist Charles Van Riper (1905–1994) wrote that the connection between forced hand switching and stuttering remained uncorroborated.[76] Still, it hadn't been disproven. Van Riper proposed that the question could be easily answered by examining a large population of all ages that had been switched compared with a control group that hadn't been switched to determine which had the greater number of stutterers (357). Yet, as Van Riper noted, such a study has not been done and, contrary to Van Riper's assumption, there are statistical and definitional issues that make undertaking such a study extremely problematic.

More than four decades later, Corballis reports, studies continue to link ambidexterity, uncertain handedness, and weak cerebral asymmetry with stuttering, learning disabilities, and mental illness. "Whether weak cerebral lateralization is the cause or the result of speech impairments," it is time, he writes, for a reexamination of the evidence for an "association between disability and failure of cerebral dominance."[77]

Afterword: What about the Zulu?

But, what about stuttering among the Zulu? Does their continued repression of left-handers translate into a higher incidence of stuttering? Although it is difficult to obtain valid data, reports indicate that stuttering is a serious problem among contemporary Zulus.[78] Outside of the West, the majority of the planet's natural left-handers are routinely forced to adopt right-handedness. And in every one of these cultures, stuttering remains a common problem of childhood. For instance, in India, where retraining left-handers is a common practice, recent data report that 7.49 percent of the population has speech disabilities, mainly stuttering, in comparison to 1 percent of the Western world.[79] Exact numbers are hard to come by, but indirect evidence suggests that stuttering in China is more frequent than in the West.[80] Even if we could uncover reliable data of higher rates of stuttering among all cultures that enforced switching left-handers to use their right hand, it would not prove a causal link to stuttering.

It would, however, provide some tantalizing evidence for the need to undertake systematic investigations.[81]

On the other hand, it is possible that the ancient and continuing negative attitudes toward left-handers reflect a real risk of mental and other disabilities. If so, why are there any left-handers at all? Why hasn't human left-handedness been bred out of existence or at least significantly reduced in numbers? Or, is there some selective advantage to left-handedness, similar to that of sickle cell disease, in which those who inherit only one allele gain protection against a more lethal harm (malaria)? Alternatively, does left-handedness arise from environmental pressure unrelated to handedness itself? Or, is there more than one route to left-handedness? Chapter 6 examines how researchers have attempted to answer these questions.

6

FROM GENES TO POPULATIONS
The Search for a Cause

There is no fixed differentiation of response in either hand until social usage begins to establish handedness.

JOHN B. WATSON (1925)

The extrinsic environment impresses circumstantial and topical configurations, but a certain nature is given, and it is for this reason that we discover such early evidences of individuality in the human infant.

GESELL AND AMES (1937)

The early twentieth-century dispute between the criminologist Cesare Lombroso and the anthropologist Robert Hertz about the causes and consequences of left-handedness was informed by their wider assumptions about the etiology of human behavior.[1] They interpreted similar evidence but did so within the frame of their own disciplinary perspectives and cultural and political agendas. Other researchers in North America at the time similarly used left-handedness as a proxy for larger philosophical and disciplinary claims. By the second decade of the twentieth century, however, this theoretical approach was increasingly challenged by researchers who were using statistical methods to examine large populations.

These later investigators insisted that population studies provided an objective approach to understanding the mysteries of human handedness, though of course their assumptions were hardly value free. Nevertheless, by midcentury the standard of practice required that investigators who wished to identify the extent and cause of human left-handedness provide quantitative evidence. By the 1980s, the assumptions that drove population studies would be challenged by a new environmental hypothesis based on a complex interaction of trauma, hormones, and autoimmunity. In this chapter we will focus on the genetic studies of the first 80 years of the twentieth century and in chapter 7 on the environmental hypothesis that challenged them.

Whether or not they agreed with Lombroso's negative characterizations of left-handers, many researchers who came after him believed
that left-handedness resulted from a combination of evolutionary and
hereditary mechanisms. The observation that left-handedness was
more frequent in children of left-handed parents persuaded many observers that handedness was inherited. Others, however, insisted that
environmental influences, including imitation and uterine trauma,
explained higher familial rates. Each view cited clinical histories as
evidence, often of patients with learning disorders. (We examined
many of these case histories in chapter 5.)[2] Finally, frustrated by the
inconclusiveness of these contradictory claims, a small number of
researchers began to explore the possibility that there were several
routes to left-handedness, some environmental and others genetic.

The ongoing debate over what constitutes left-handedness, or
even at what age it could reliably be determined, focused in part on
whether left-handedness was defined by *preference* or *skill*, with preference being fixed at birth or earlier, and skill around age 3. Do handedness studies count as right-handed or left-handed those who have
been "retrained"? Do researchers always know if their subjects have
been switched; do the subjects themselves? When subjects are asked
to report whether their parents and grandparents are right- or left-
handed, would the subjects know if their relatives' hand preference
had been switched? And even if they knew, how should those relatives
who had been switched to right-handedness be classified? Finally, is
there a selective advantage for left-handedness that explains its persistence either as preference or as skill?[3] To confront these questions,
we return to the studies of the early twentieth century and then to the
quantitative focus of the studies that followed.

THE EYES HAVE IT

In a series of articles and in a widely discussed 1908 book, ophthalmologist George Milbry Gould argued that handedness was determined by whether a person favored his or her right or left eye.[4] A
respected and influential physician, Gould served as president of
the National Medical Library Association and editor of the *Philadelphia Medical Journal* and *American Medicine and Medical News.*
Gould also compiled and published extensive medical dictionaries.

George M. Gould (1848–1922), ophthalmologist, president of the National Medical Library Association, and editor of the *Philadelphia Medical Journal* and *American Medicine and Medical News*. Gould published articles and a 1908 book arguing that handedness was determined by whether a person favored his or her right or left eye.

His investigations of medical anomalies had a wide readership among professional and popular audiences. In 1896 Gould joined with the equally respected ophthalmic surgeon Walter Lytle Pyle, who was, among other honors, a fellow of the American Academy of Medicine, to publish *Anomalies and Curiosities of Medicine*. Their book, still in print, aimed to provide scientific explanations for physical and mental deviations that the then popular press, circuses, and carnival displays labeled as freaks.[5]

Beginning in 1904, in a series of medical articles and his popular book in 1908, Gould turned his attention to another medical anomaly, left-handedness. Based on both his clinical experience and his distaste for magical thinking, Gould argued that handedness was determined by a dominant eye. In most humans, wrote Gould, the right eye was stronger than the left: a child grasps with his or her right hand because the right eye is visually and anatomically superior to the left.[6] According to Gould right-eye dominance rather than handedness was inherited. An infant, wrote Gould, will grasp objects with the left hand if the left eye sees better than the right (601). Therefore, Gould concluded, while left-handedness itself was not hereditary, the dominant eye was.

Gould's theory was widely and favorably reported in the medical and popular press. The *New York Times* ran a comprehensive article entitled "The Riddle of Right-Handedness Made Clear," in which Gould was credited with having "found the answer" to the long-debated question of the origin of left-handedness.[7] Similar praise, along with detailed reports, appeared nationally, including articles in *Collier's* magazine[8] and in the press as far away as Perth, Australia.[9]

Many geneticists were skeptical. The common finding that two right-handed parents could have left-handed offspring suggested that a recessive gene was responsible. Writing in the *Journal of Heredity* in 1911, Harvey E. Jordan (whose views on retraining were elaborated in chapter 5), dismissed Gould's rejection of the heritability of handedness. It made more sense, wrote Jordan, to assume that both eye and hand dominance resulted from the same cause.[10] Jordan shared other eugenics beliefs—that "human stock" could be improved by selective reduction of "socially defective" individuals through a variety of means, including compulsory sterilization. For Jordan, Mendelian inheritance provided the scientific validation for eugenic theory and policy.[11]

It was from this eugenic perspective that Jordan agreed to write a foreword to American writer Beaufort Sims Parson's best-selling 1924 book, *Left-Handedness: A New Interpretation*, less because he was persuaded by Parson's hypothesis than because the study provided a popular presentation of Mendelian mechanisms for the cause of left-handedness. Jordan also endorsed Parson's argument because Parson, who was not a physician, relied extensively on Jordan's publications and endorsed Jordan's assertion that left-handedness resulted from a Mendelian recessive gene.[12] "When a left-handed individual marries a right-handed individual," wrote Jordan in 1922, "the children from such a marriage are all right-handed, due to the fact that right-handedness dominates in heredity." When two of these "right-handed" but "hybrid . . . offspring intermarry, however, the children are right-handed and left-handed approximately in the proportion of three to one." If two left-handers marry, all their children will "show a bias toward left-handedness" (381).

Parson theorized that nonhuman primates and early humans saw with either eye as needed. Both eyes worked together (binocular vision) to produce a single image, and there was no dominant human left or right eye. Modern human vision, wrote Parson, evolved from this primitive binocular vision to what Parson labeled "unilateral sighting," in which one eye, mainly the right, was dominant. Like

Jordan, Parson endorsed Lombroso's assertion that early humans and "primitive races" were ambidextrous.[13] While "binocular vision accompanied by ambidexterity" represented "the ancient parent type," modern humans were distinguished by a mutation that resulted in the combination of right-eye dominance and right-handedness.[14]

The dominance of the right eye and the right hand was influenced by environmental factors such as the purported advantage of carrying weapons in the right hand. This behavior, Parson hypothesized, was intensified by the use of the shield held in the left hand in order to protect the heart and other left-sided vital organs. Parson also implicated other influences, including sun worship, gesture, and finger counting. As a result of these evolutionary pressures, the ancient ambidextrous/binocular vision gene, which manifested itself in left-handedness, existed as a regressive variant, that is, a Mendelian mutation.[15] Parson invented a device he called a "manuscope" that purportedly measured the strength of the relationship between handedness and eyedness. The manuscope itself consisted "of a small darkened box or camera-like chamber, one end of which fits over the eyes and upper part of the face after the manner of a stereoscope, from whence it tapers for 9 inches to a circular aperture 1⅛ inches in diameter at the farther end. Moveable shutters on each side of the median line permit the instantaneous blocking of the left and right lines of sight" (78).

Parson's book received some positive reviews,[16] but aside from Jordan, most were skeptical of Parson's hypothesis. Writing in the *Journal of Abnormal and Social Psychology*, psychologist June E. Downey found that Parson's theory failed to explain the cause of right-eye domination.[17] According to a reviewer for the *British Medical Journal*, Parson, who was not an ophthalmologist, offered a "flawed" theory not supported by clinical experience. "Every ophthalmic surgeon," the reviewer wrote, "has seen patients who are right-handed with a dominant left eye."[18] Parson's theory was unpersuasive, wrote University of Washington professor Stevenson Smith in the journal *Science*, because it could not explain the existence of a congenitally blind left-handed infant.[19] Along with Gould's theories of the connection between eye dominance and handedness, Parson's hypothesis faded from handedness literature. Researchers today have gone even further, arguing that "lateralization of verbal and visuospatial functions in the same hemisphere" were unrelated, resulting from "independent developmental factors."[20] But both serve as examples of the extent to which focus on one trait (eye dominance) was employed to explain another trait (handedness). From that perspective both Gould's and Parson's studies are similar to current investigations of the etiology of left-handedness.

In chapter 1 we examined a recent study using the latest gene identification technology (GWAS) that identified the gene for handedness as *PCSK6*, an enzyme that codes for left and right asymmetry in a wide array of species. According to the study's authors, disruption of this allele interferes with normal asymmetric development in humans, resulting in human left-handers.[21] Like Gould a century earlier, these researchers were hailed by the media as having discovered the genes that determined human handedness.[22] Current geneticists questioned the study because it had not been replicated or even tested on a normal control population,[23] and like Jordan's reaction a century ago, a recent announcement connecting left forelimb preference among tree kangaroos (marsupials) to their bipedalism, and by extension laterality in bipedal humans (mammalians), received wide and positive media coverage.[24] These red-necked wallabies lack a corpus callosum. Combined with their left-forelimb preference, this has suggested to some a connection to human autism, because the corpus callosum has been reported to be underdeveloped in some people with autism, who are putatively more likely to be left-handed.[25] Most researchers find this too big a jump to make. Over and over we see that it is much easier to uncover the flaws and limitations in older investigations than it is to identify them in current claims, in great part because each generation shares a common scientific vocabulary.

THE RISE OF POPULATION STUDIES

Jordan endorsed Parson's book, but he led the movement away from speculative handedness theory to quantitative population studies. His 1922 survey of 2,762 university and public school pupils provided a model that others would follow.[26] From this larger group Jordan selected 22 "left-handed families," from whom he collected extensive pedigrees. Conceding that his methods and limited data could not conclusively prove his Mendelian hypothesis, he nevertheless found it significant that none of his sample contradicted his hypothesis.[27]

Jordan urged others to examine larger populations to validate a Mendelian mechanism.

Responding to Jordan's call for larger population studies, University of Colorado professor and ecologist Francis Ramaley* interviewed 610 parents and their 1,130 college student children. As Jordan had hoped, Ramaley reported that left-handedness was a recessive (Mendelian) trait, because it could be found in children with no left-handed parents—sometimes in children with no left-handers for several generations. According to Ramaley's data, one-sixth of the population is left-handed.[28]

Ramaley's paper caught the attention of Ohio State University geneticist Herbert D. Chamberlain, who decided to test Ramaley's hypothesis that left-handedness was a recessive trait. One of Chamberlain's reservations about Ramaley's study was that his population was not large enough to support his hypothesis. To overcome this limitation, Chamberlain and his colleagues interviewed 2,177 students of the incoming 1927 Ohio State University freshman class. The students were asked to state their handedness and that of their parents and siblings. Chamberlain's total population, including parents and siblings was 12,066, 10 times greater than the Ramaley study.[29]

The Chamberlain group reported that 4.31 percent of the students were left-handed. This percentage, remarked Chamberlain, was similar to those reported by others. (Of the 4,354 parents 4.13% of the fathers were left-handed, while only 2.94% of the mothers were left-handed; 3.76% of the sisters were left-handed, as were 6.05% of the brothers.) Of the entire population of over 12,000, 4.34 percent were left-handed. Chamberlain initially had hoped his large population would demonstrate that left-handedness was recessive, but he was unable to do so because his team could not control for the large number of variables they uncovered. While they could not demonstrate that left-handedness was a Mendelian recessive trait, the Chamberlain team insisted that their data did show that left-handedness was inherited.[30]

Of course, these early twentieth-century geneticists lacked the sophisticated statistical tools, such as modern regression analysis, that

---

*Ramaley was a Fellow of the American Association for the Advancement of Science and member of the American Society Naturalists, Botanical Society America, Ecological Society America, Society Experimental Biology and Medicine.

we take for granted. Only in 1922 did R. A. Fisher develop a method for applying regression analysis,[31] and not until the late 1940s did he fully refine his models in a way that would have been useful to Chamberlain.[32]

Like Chamberlain, most experts continued to focus on the genotypical evidence. Thus, the renowned pediatric psychologist and director of the Yale Clinic of Child Development, Arnold Gesell concluded that because handedness was one of the "fundamental" ways that children "assert themselves under varying environmental conditions," it must be inherited.[33] This view was endorsed by educational psychologist K. C. Garrison, who repeated, as many others had reported, that because two right-handed parents could give birth to left-handed offspring, that the mechanism for left-handedness was probably recessive, but certainly inherited.[34]

Nurture and the Case History

Not everyone, however, was persuaded that left-handedness was an inherited trait. Most vocal among the dissenters was the founder of American behaviorism, the social determinist psychologist John Broadus Watson (1878–1958). Watson claimed that handedness resulted from social and parental pressures.[35] Watson insisted that there was "no fixed differentiation of response in either hand until social usage begins to establish handedness." Soon after, wrote Watson, "society . . . steps in and says 'Thou shalt use thy right hand.' Pressure promptly begins. 'Shake hands with your right hand, Willy.' We hold the infant so that it will wave 'bye bye' with the right hand. We force it to eat with the right hand. This in itself is a potent enough conditioning factor to account for handedness. The main problem is, I believe, settled: handedness is not an 'instinct.' It is possibly not even structurally determined. It is socially conditioned."[36]

Watson's influence on academic psychology grew over the next decades, and his conclusions, though not always his behaviorist theories, were endorsed by a number of educators. Others, attracted by psychoanalysis, attempted to merge Watson's views about handedness with psychoanalytic theory.

As we saw in chapter 5, those who advocated hand switching were the strongest advocates of the nurture hypothesis. They focused on case histories rather than population studies. Many of them were psychoanalytically oriented, attaching left-handedness to defective parenting, defiant personality disorders, and childhood neuroses in general. They too insisted that left-handedness resulted from a combination of permissive education and lack of parental discipline. These factors supposedly encouraged oppositional behavior, disobedience, and rebelliousness that manifested itself in a refusal to use the right hand.[37] Chief among this group was New York University pediatric psychiatrist and former psychiatrist of the New York City Board of Education, Abram Blau, whose views were elaborated in chapter 5. For Blau, writing in 1946, handedness was, "an acquired cultural trait and not an inherited one."[38]

Toward a Mixed Model

A few observers, skeptical about the nature-versus-nurture dichotomy, suggested that there were distinct and different causes of left-handedness.[39] In a 1936 monograph sponsored by the University of Chicago Committee of Child Development, Minnie Giesecke reported that she had followed 17 infants for between 2 and 17 months to determine if handedness were inherited or taught. Giesecke noted that the researchers' conclusion was influenced by their presumption of the age when handedness was determined. But a careful examination of the literature revealed that there was no agreement about when children chose their dominant hand.[40] To resolve this question Giesecke followed three sets of infants ranging in age from birth to 17 months. Her observations revealed that hand and side preference could solidify at any time during the first 17 months, depending on a combination of psychological factors and practice and training in manual skills. Summing up her findings she reported that "hand preference . . . may have a physiological basis in the tendency toward greater activity in one side of the body, and also a psychological basis in the greater amount of practice and training received by the hand of the dominant side due to its more frequent use in the early development of manual skills" (83). Giesecke acknowledged a possible role of eye dominance in determining side preference. Although she conceded a physiological role in hand selection, Giesecke found no evidence of hereditary influence in this scenario. Moreover, case histories revealed that while

35 percent were left-dominant as infants, the percentage declined as the children matured. Thus Giesecke endorsed the hypothesis of the German educator M. Ludwig[41] that there might be "two types of left-handedness, genotypical and phenotypical."[42]

Building on these studies, in her widely cited 1957 book, *Left-Handedness: Laterality Characteristics and Their Educational Implications,* Scottish educational psychologist Margaret M. Clark endorsed a modified mixed model. In contrast to Clark, Giesecke concluded that handedness was generally hereditary, because the chances of being left-handed increased when left-handedness could be found in a family. But, like Giesecke and Ludwig, Clark insisted that nongenetic factors played an important role in determining whether a particular child will be right- or left-handed. These included where one lived, local attitudes toward left-handers, and the tension between temperament and environmental pressures. Nevertheless, Clark assumed that while changes in attitudes and practices enabled many formerly suppressed left-handers to emerge, the underlying causes of handed preference were genetic.[43]

THE 1970S AND 1980S

Despite being widely cited, Clark's book, especially her suggestion that some handedness might be determined by a combination of both preference and skill, largely has not been followed up. In fact, genetically inclined psychologists often cite Clark's work as evidence of the genetic origins of left-handedness. A University of Colorado research team composed of psychologist Jerre Levy and physicist Thomas Nagylaki proposed a two-gene, four-allele mathematical model with one gene coded for speech dominance in either hemisphere and the other for hand preference which could be either "contralateral or ipsilateral" in relation to the dominant hemisphere. The dominant allele combined left-hemisphere speech with right-handed preference.[44]

The hypothesis was criticized by a number of experts, beginning with British psychologist Patrick Hudson, who argued that Levy and Nagylaki's model failed to fit the data, especially the findings that males were more likely to be left-handed than females and that there were generational differences in the prevalence of left-handedness.[45] British psychologist Marian Annett, who was developing her own mathematical hypothesis of handedness (discussed at length later), provided a more extensive critique of Levy and Nagylaki's interpretation of their data, especially their assumption that left-handers were generally right-brained.[46] Finally, University of Hawaii geneticist G. C. Ashton and colleagues were skeptical not only of Levy and Nagylaki's model but also of two new genetic models (discussed later): one proposed by Annett and the other by British neurosurgeon and psychologist Chris McManus. Ashton and colleagues characterized both of these models as "notable more for the ingenuity of their alleged fit to an investigator's data than for their experimental verification."[47]

Insisting that a purely genetic model was unsupported by the data, in 1981 Ashton and colleagues proposed multiple causes for left-handedness. Based on an extensive review of the studies discussed in this chapter and their examination of 1,818 Hawaiian nuclear families, the Ashton team hypothesized that "left-handedness results from a combination of genotype, birth experience, and maternal example." Their study focused on the interaction of four factors. The first, a general "genetic factor," accounted for approximately 20 percent of hand preference variation, but they conceded "the nature of the genetic factor could not be determined." The second was "a birth stress factor," as described by Bakan, Dibb, and Reed[48] and by Coren and Porac.[49] The third included maternal influences, which were responsible for an undetermined but significant proportion of handedness variation. Ashton and colleagues could not determine whether this factor was due to imitation or genetic inheritance.

These three factors produced "additive effects." A fourth factor was aging, which was purported to reduce the number of left-handers. Their evidence for this factor came from the decline of left-handers as the population aged, which, they hypothesized, resulted as much from cultural influences that reversed left-handedness. But Ashton and colleagues admitted that they had no evidence to support this hypothesis.[50]

It is not surprising that the four-part hypothesis had little influence on the competing theories that emerged in the 1980s and that would be refined in the three decades that followed: the research findings were vague, and the researchers could not validate the role,

influence, and interactions of these four factors. Thus, attempts to lo-
cate a selective advantage for the persistence of left-handers continue
to intrigue researchers.

Although Clark's work is widely *cited*, it has not always been carefully
read. Only Ashton and his colleagues paid attention to Clark's call for
a multifactorial research agenda. To be fair, Clark ultimately came
down on the side of nature, and her suggestion of a role for nurture
was deeply tentative and not fully elaborated.

Multifactorial investigations of left-handedness essentially disappeared once the Ashton hypothesis had been rejected. Why? One
reason was the renewal in the 1980s of the debate between nature
or nurture. The case for an environmental hypothesis emerged in
the investigations of Harvard neurologist Norman Geschwind and
his colleagues, in the early 1980s, which examined the connections
between uterine trauma, immune disorders, hormones, developmental disorders, and handedness.[51] Simultaneous with Geschwind's
hypothesis, others developed elaborate genetic mathematical models for left-handedness. Their creators, especially Marian Annett and
Chris McManus, were deeply hostile to the Geschwind hypothesis (and
sometimes to each other), viewing Geschwind's claims as both a threat
to and a diversion from identifying the etiology of handedness. We
will examine Geschwind's hypothesis in chapter 7 and then return to
his critics and their genetic models in chapter 8.

7

THE GESCHWIND HYPOTHESIS

In my opinion, Norman Geschwind ranks as the single most influential figure of the 20th century on the study of aphasia; [his] crowning achievement was to return the brain to an important place in the investigation of language functions.

FRANK BENSON (1997)

The left brain / right brain bandwagon had brought the laterality field into disrepute with most serious psychologists. What little credibility remained has been swept away by the rush to follow the GBG model.

MARIAN ANNETT (1994)

In 1973 Canadian psychologist Paul Bakan and colleagues reported that ambidextrous and left-handed responders were twice as likely as right-handers to report that there had been complications at their births.[1] Based on this report and subsequent studies, Bakan concluded that left-handedness resulted from insults in utero or at birth.[2] Reading this, I wondered if my left-handedness could have similar roots. My mother was pregnant with me during an extremely stressful period of her life. My father, his brother, and my mother's two brothers were fighting in the Pacific and Europe. My mother's experience was hardly unique. I am intrigued by the possibility that the incidence of left-handers might have been higher among children born during the Second World War, though as far as I know this study has not been done. Indeed, the effect should be seen worldwide.

Psychologist and neurosurgeon Chris McManus argued that the evidence failed to support Bakan's hypothesis. Examining two large retrospective studies and one larger prospective study, McManus concluded that there was no statistical or other persuasive connection between left-handedness and birth stress.[3]

In 1981 Harvard University neurology professor Norman Geschwind proposed a more complex connection between uterine stress and left-handedness. Along with his colleagues Peter O. Behan and Albert M. Galaburda, Geschwind hypothesized that uterine stress boosted the transmission of testosterone, increasing the risk for allergies, immune disorders, learning disabilities, and left-handedness.[4] The Geschwind hypothesis (called GBG, for Geschwind, Behan, and Galaburda) would reframe the debate about the etiology of left-handedness for the next two decades.

Geschwind brought enormous credibility to his hypothesis. Author of the 1965 landmark study *Behavioral Neurology*, Geschwind had a reputation as a brilliant and passionate intellectual who continually imagined connections and associations that others had missed. His "approach did not reflect a single flash of insight," wrote his colleague Thomas D. Sabin, but it represented "an accumulation of knowledge in astounding detail, in which Norman saw elaborate and profound relationships."[5] Harvard psychologist Howard Gardner referred to Geschwind as a "creative genius." Gardner reported that Geschwind "could discuss complex issues in a straightforward way, bring new excitement to a discussion, and respond to even the most vexing questions with insight, appropriateness, and timely wit": "Nearly everyone who heard Norman Geschwind concurred that he was one of the great scientific lecturers of the age." His intellectual rigor "exerted a powerful influence even on those who staunchly disagreed with his claims."[6]

Geschwind's childhood and family experiences influenced his skepticism of psychiatric and neurological dogmas. In particular his yeshiva education, with its insistence that there could be no final reading of even a sacred text, sanctioned Geschwind's later challenge of medical authorities.[7]

His interests in handedness and learning disabilities can be traced to his 1965 study of "disconnection syndromes."[8] Geschwind was skeptical of the insistence of many of his psychiatry professors that behavior was unrelated to neuropathology. From this perspective the science and practice of neurology were restricted to the diagnosis and treatment of aphasias and epilepsies, while cognitive impairments and developmental disorders were classified as functional (psychological) disorders. To the contrary, Geschwind found that many of the patients he examined with neurological deficits also exhibited behavioral (developmental) disorders.[9]

Norman Geschwind (1926–1984), Harvard University neurology professor, who in the early 1980s, along with his colleagues Peter O. Behan and Albert M. Galaburda, hypothesized that learning disabilities, allergies, immune disorders, talent, and left-handedness were connected to uterine stress. (Photo courtesy of Kenneth M. Heilman.)

Albert M. Galaburda, collaborator with Norman Geschwind on refining the so-called GBG hypothesis connecting uterine stress with a variety of developmental and immune disorders and with atypical brain lateralization. Later in his career Galaburda would revise his thinking about the hypothesis. (Photo courtesy of Albert Galaburda.)

Geschwind's generation also had been taught that aphasias resulted from global rather than localized or focal neurological lesions. Adherents to this so-called holism were dismissive of the work of nineteenth-century German aphasiologist Carl Wernicke (1848–1905), suggesting that Wernicke's localization was an updated version of phrenology.[10] Unwilling to accept secondhand characterization of Wernicke's writings (or of any text for that matter), Geschwind reread Wernicke in the original German. Much to his surprise Geschwind discovered a multilayered explanation in which Wernicke distinguished between localized lesions that affected motor movement and multiple lesions located in association pathways, which, when extensive, resulted in behavioral disorders.

He then examined the works of the holists, discovering that while their rhetoric rejected Wernicke, their explanations of aphasias reinforced Wernicke's hypothesis. Building on his interpretation of these historical documents and on his clinical experiences, the then 35-year-old Geschwind urged that Wernicke's disconnection syndromes, which he labeled "Behavioral Neurology" be resurrected and expanded.[11]

Going further than Wernicke, Geschwind suggested that higher-function disorders resulted from disconnection.[12] Examining the effect of aphasias, Geschwind, who was right-handed, speculated that often, but not always, right-handers and left-handers had different cognitive presentations despite having similarly located lesions. For example, aphasia in a left-hemisphere dominant person may manifest as depression, while aphasia in a right-hemisphere dominant person may manifest as euphoria.[13] He noted that the language area of the brain (the planum temporale) was larger on the left hemisphere in the vast majority of the population.

Geschwind also observed that the left hemisphere normally matured more slowly than the right, allowing the speech areas to develop with less possibility of injury. But in some fetuses the opposite

Stanley Coren, University of British Columbia psychologist and author of the influential and widely read 1992 book, *The Left-Hander Syndrome: The Causes and Consequences of Left-Handedness,* which cited the GBG hypothesis as support for his claim of a causative relationship between mental disorders and left-handedness (see p. 104). (Photo courtesy of SC Psychological Enterprises LTD.)

was true: more rapid development of the left hemisphere exposed it to greater risk of damage.[14] Geschwind believed that examining the interaction of these mechanisms could open new approaches to understanding disease processes, including such seemingly diverse phenomena as language recovery after an aphasia, developmental disorders, dementias, and, counterintuitively, human talents.[15]

This was the context in which Geschwind, Behan, and Galaburda hypothesized that learning disabilities, allergies, immune disorders, talent, and left-handedness were, like complex aphasias, association disorders with a common mechanism but with different presentations—which were determined by the nature and timing of environment insults in utero. GBG argued that uterine stress increased transmission of testosterone, damaging the left hemisphere. This reversed the normal right-hand and left-hemisphere linguistic dominance in favor of the left hand and right hemisphere. The effect of amplified testosterone transmission was greater in males than in females because males already had higher testosterone secretions due to their fetal testes. Thus males were more likely to be left-handed and linguistically right-brained than females. The increased testosterone secretion also interfered with neuronal migration and slowed maturation of the thymus, which is charged with recognizing self and nonself antigens. This process, wrote Geschwind, contributes to an increase in autoimmunity later in life.[16]

RAISING THE EVIDENCE BAR

Geschwind developed his hypotheses from clinical experience augmented by his deep knowledge of neuropathology. He was profoundly skeptical of population studies of neurological illnesses that were not grounded in neurological mechanisms. Although they were built on increasingly sophisticated statistical methods, most studies, Geschwind found, were unreliable; that is, their variables might have the same name, but they were not necessarily the same thing. This was especially true, as we discussed in chapter 3, when it came to defining left-handedness, because different studies defined left-handedness in different ways or were overly inclusive, failing to distinguish between people with strong and mild hand preferences.[17] It made no sense to

Geschwind to conduct population studies on handedness or immunity if researchers could not agree on a common definition of the data
and variables to be included.

He was also troubled that many of the authors of population studies were not conversant with the neuropathology. As one of his colleagues recalled, in his classes Geschwind would have the students
read flawed articles to make sure that they could identify "faulty arguments, especially misleading statistical analyses." Geschwind also
tied the limitations of population studies to the lack of reliability of
data—such as assuming that all left-handers could serve as a proxy
for linguistically right-brained individuals or that every diagnosis of a
particular learning disorder or immune illness had the same etiology
and course. He noted that reports of the prevalence of neurological
disorders varied depending on whether the diagnosis was made by
internists, psychiatrists, or neurologists. Different specialists often
used different criteria in making neurological diagnoses. Because of
their diagnostic assumptions, researchers not trained in neurology
often failed to identify neurologic illnesses, and as a result the incidence of neurologic disorders was underreported or misreported.[18]
His colleague, neurologist Thomas Sabin, wrote that Geschwind was
convinced that closely examining a single case intensively was more
productive than a statistical analysis of a large number of inadequately studied individual patients. Geschwind, recalled Sabin, "was
fond of citing many instances in which a single case report had resulted in an incremental step forward in neurology."[19]

It is ironic, then, that the GBG hypothesis became popularized and
disseminated based mainly on three population studies conducted
by Geschwind and his neurology colleagues Behan and Galaburda in
the early 1980s. Geschwind believed their investigations overcame the
problems of other population studies, reinforcing his clinical observations while, he hoped, forcing researchers to turn their attention to
environmental dynamics.[20]

Geschwind and Behan conducted two population studies in 1981
in London and Glasgow.[21] The first study was divided into two parts.
Part 1, conducted by Geschwind and Behan in 1981–82, compared
500 strong left-handers with 900 strong right-handers.[22] Seeking
reliability, GBG restricted the subjects they recruited for their studies
to those whose score on the Oldfield/Edinburgh Handedness Inventory was either +100 (right-handed for all tasks) or –100 (left-handed

for all tasks). This part of the study reported that left-handers had 2.7 times more immune disorders and 10 times more learning disabilities than right-handers. The left-handers also reported significantly higher numbers of first- and second-degree relatives with immune disorders.[23]

In the second part of this study, Geschwind and Behan compared a population diagnosed with immune disorders and normal population of left- and right-handers. Again, opting for reliability, they restricted the population of those with an immune disorder to those whose disorder had been diagnosed in a teaching hospital. The subjects included 247 left-handers and 647 right-handers. The relative rates were similar to those reported in part 1 of their study, with left-handers 2.3 times more likely than right-handers to report immune disorders in themselves and their relatives. Left-handers and their relatives again were found to have a much higher rate of learning disorders than the right-handers.[24]

Of the 1,396 subjects in both parts of this study, 652 (ages 12 to 50) were strongly right-handed, as determined by an Oldfield Inventory score of +100. Of this group, 278 were male and 378 were female. Of the 440 strongly left-handed subjects (ages 11–50), again based on a –100 of the Oldfield Inventory, there were 180 males and 260 females. Of the 340 patients with diagnosed autoimmune disease (ages 16–54), 130 were males and 174 were females. Aside from those recruited at the Anything Left-Handed Shop in London, all subjects were recruited in Glasgow. In addition to their filling out the Oldfield Handedness Inventory, subjects were asked a series of questions, by a single investigator, about their medical history and that of their families.

Geschwind and Behan designed a second study aimed at capturing the extent of left-handedness in people diagnosed with immune disorders or migraines compared to a control population.[25] This study consisted of 142 patients diagnosed with migraines, 98 patients diagnosed with myasthenia gravis, and 1,142 controls from the general population of Glasgow. They found a statistically significant percentage of left-handers among patients diagnosed with severe migraines and among patients diagnosed with myasthenia gravis. However, they reported no significant frequency of left-handedness among those diagnosed with rheumatoid arthritis, mixed-collagen vascular diseases, and multiple sclerosis (5098).

Constantly reexamining their hypothesis, in 1984 Geschwind and Behan published revisions of their 1981 and 1982 studies. They now emphasized that immune disorders did not cause learning disabilities but that immune disorders shared "some common factor" that also impacted the frequency of left-handedness, learning disabilities, and autoimmunity.[26] Geschwind and Behan had warned that those at risk were not necessarily left-handed; at risk, instead, were those with *anomalous dominance* (AD), where normal left-hemispheric language laterality shifted to the right hemisphere. Nevertheless GBG's population studies had overrepresented left-handers. That was because up to 30 percent of left-handers are either right-brained or bilateral for language, while only 5 percent of right-handers appear to be right-hemisphere dominant. Thus there were 3 out of 10 chances of recruiting a right-brain dominant subject among left-handers, but only a 5 per 100 chance in a population of right-handers. Because the right-handed populace was ten times greater than that of left-handers, a much larger population of right-handers than left-handers had to be examined in order to capture significant numbers of right-brained, right-handed subjects. As a result left-handers were more often selected in studies seeking to enroll right-hemisphere dominant subjects. This is why GBG concluded that "handedness is a useful but potentially treacherous marker in as much as many of those with anomalous dominance are right-handed."[27] "It is easy to be misled," warned GBG, "by the fact that the studies reported here have compared strongly left-handed and strongly right-handed people, the strong sinistrals being found to have a much higher rate of immune disease and learning disabilities than the dextrals."[28]

Geschwind died in 1984 at age 58. During his final years he joined with Galaburda in revising their understanding of neuronal and biological mechanisms of the cerebral dominance hypothesis. The result, *Cerebral Lateralization: Biological Mechanisms, Associations, and Pathology*, was published in three issues of *Archives of Neurology* and issued as a single volume in 1987.[29] *Cerebral Lateralization* is a tour de force of hypothesis building, often brilliant and always creative. It also is highly speculative, densely written, a bit redundant, and, most of all, a work in progress.

The credibility of the GBG hypothesis, Geschwind believed, ultimately rested on understanding the neuropathological mechanisms. Thus, along with Galaburda, he reviewed, often in excruciating detail, human cerebral asymmetry, standard and anomalous hemispheric dominance, hormone actions, genetics and epigenetics, learning disabilities, and immunity. They examined other conditions, including neural crest disorders, immunity and infection, brain disease and laterality, pharmacology, metabolism, cyclic phenomena, and the origins of asymmetry (156, 197).*

Having concluded that left-handedness was not a useful or valid proxy for identification of the neuropathology and autoimmunity that the researchers were seeking to explain, they focused instead on anomalous dominance, which could be found in both left-handers and right-handers. Thus, while interest in the GBG hypothesis had been sustained because it seemed to provide an environmental explanation of left-handedness, Geschwind's final study concluded that left-handedness was neither a cause nor an effect of the disorders that he and Galaburda sought to explain. Because non-right-handers are more likely to be categorized as ambidextrous, Geschwind and Galaburda hypothesized that they would also have a higher rate of bilateral linguistic representation (79).

From this perspective the researchers concluded that individuals should be classified by hemispheric dominance rather than by right-handedness or left-handedness. They regretted their earlier reliance on the Oldfield Handedness Inventory, because whatever their utility, handedness inventories could not identify hemispheric dominance (73).

Geschwind and Galaburda insisted that their hypothesis was a work in progress, the starting point for future research.[30] Despite their earlier population studies, Geschwind and Galaburda based their conclusions on clinical experience and neurobiological knowledge. Today, the research gold standard is the double-blinded large clinical trial.[31] From this perspective Geschwind's insistence on a patient-based investigation and current scientific knowledge is not only out of style but also cannot be funded, which is the true gold standard in medical research.

---

* The page references here are from the book version. N. Geschwind and A. M. Galaburda, *Cerebral Lateralization: Biological Mechanisms, Associations, and Pathology* (Cambridge, Mass.: MIT Press, 1987).

GBG MEETS FANS AND CRITICS

The GBG hypothesis received an enthusiastic response from the neuropsychiatric community and extensive coverage in popular science media, such as the *New York Times* and *Psychology Today*.[32] The influence of the GBG hypothesis is evident from the huge number of citations, almost 600, to GBG's papers from January 1986 to March 1993. It was obvious, to the authors of a 1994 study, that GBG was "a major scientific phenomenon, of sudden onset and immediate influence."[33] Geschwind, they wrote, presented wide "empirical support from a dozen or more subdisciplines, and seemed to have something important to say about almost every aspect of neuropsychology."[34]

The Geschwind hypothesis was cited as providing powerful scientific evidence in support of a pathological interpretation of left-handedness, despite Geschwind and Galaburda's rejection of such a connection. Many expected, again despite Geschwind's skepticism, that new population studies would validate Geschwind's clinical insights. In the next decades hundreds of studies were published claiming to substantiate the GBG hypothesis. Most of these investigations attributed to Geschwind the claim that uterine trauma caused a variety of learning and immune disorders, left-handedness, and right-hemispheric dominance. Given this interpretation of a common pathophysiology, left-handedness continued to serve as a proxy for all these disorders despite GBG's insistence that anomalous dominance, which was not restricted to left-handedness, provided the key to learning disabilities and human talents.[35]

The original GBG hypothesis was translated to the general public in accessible language in a variety of popular books and articles.[36] The most influential and widely read of these was University of British Columbia psychologist Stanley Coren's 1992 book, *The Left-Hander Syndrome: The Causes and Consequences of LeftHandedness*. Coren wrote of a causative relationship between in utero stress and left-handedness.[37] He reported that left-handers were more likely to develop immune disorders. In addition Coren and his colleagues made the provocative claim that left-handers experienced significantly reduced longevity compared to right-handers.[38] (See photo on page 94.)

Assuming, along with Lombroso, that left-handedness was abnormal, Coren characterized it as "right-handedness run amok." "If all human beings were genetically targeted to be right-handed," asked

Coren, "why do we have left-handers?" Answering his own question, he suggested that "left-handedness should be looked upon as a failure to reach right-handedness. If right-handedness is encoded in every human being's genetic makeup, doesn't it mean that something must have gone wrong to produce left-handers?"[39]

Endorsing GBG but citing his own published collaborative research, Coren claimed that "abnormal development of the left-hemisphere results in a shift toward left- or mixed-handedness," which he connected to the increased risk for a number of "psychological problems, including dyslexia, attention deficit disorders, learning disabilities, and mental retardation."[40] Coren labeled these associations "left-hander syndrome," thus the title of his book. Citing GBG for corroboration, Coren concluded that left-handedness resulted from injuries to the left hemisphere (140).

Coren's work helped popularize Geschwind's hypotheses. But Coren made it seem that the GBG hypothesis was aimed at explaining the etiology of left-handedness, whereas Geschwind and colleagues had, in fact, concluded that learning disabilities and talent were connected to anomalous hemispheric dominance rather than left-handedness per se. Indeed, Geschwind and Galaburda's 1985 *Archives of Neurology* articles and 1987 book had gone further, concluding that left-handedness was a misleading proxy for identifying anomalous dominance.[41]

CRITIQUING THE GBG HYPOTHESIS

Much to psychologist Marian Annett and psychologist/neurosurgeon Chris McManus's displeasure, their genetic models were overshadowed by the widespread professional and popular attention given to the Geschwind hypothesis. Although McManus and Annett were, and remain, rivals, they have been united in their opposition to the Geschwind hypothesis. A third and crucial player in these discussions is New Zealand psychologist Michael Corballis, whose contributions are examined at length in chapter 9. Because of their influence, the Annett and McManus formulations are the focus here.

The stakes were high; if the GBG hypothesis were allowed to stand, the genetic models would be seen as irrelevant. Indeed, both Annett and McManus believed that the Geschwind hypothesis not only was incorrect but also trivialized serious neurobiological examinations of handedness. Thus, they set out to expose what they believed to be its fundamental flaws.

In an extensive critique published in 1991, McManus and Canadian psychologist Michael Bryden asserted that the GBG hypothesis relied on a variety of speculations and claims that were ambiguous, confusing, and ultimately untestable. They went further: the GBG hypothesis attempted to "explain everything but at the price of being unfalsifiable."[42] In a second study published that year of hand preferences and immune disorders of 734 Canadian undergraduates, they found "no support for the argument that handedness and immune disorders are related."[43]

McManus and Bryden were distressed by the persistence and influence of the GBG hypothesis even in the face of their critiques, which they believed were persuasive. Thus, in 1994, along with Canadian psychologist Barbara Bulman-Fleming, they invited a group of handedness researchers, including Annett, to respond to a comprehensive 65-page critique of the Geschwind hypothesis.[44] The responses, consisting of 21 articles, ending with a reply by McManus and Bryden, were published in a special issue of *Brain and Cognition*.[45]

Evaluating the GBG hypothesis, McManus and Bulman-Fleming wrote, was "more akin to an exercise in hermeneutics than in conventional scientific exegesis."[46] They examined five broad areas of interrelations proposed by Geschwind, including anomalous dominance, immune disorders, learning disabilities, creativity, and neural crest disorders. They conceded that there was evidence of a "fairly strong association between AD and language disorders, with the incidence of some form of atypical lateralization being more common in developmental dyslexia, autism, and stuttering." However, the Bryden team dismissed the concept of AD as problematic, because its definition was unclear. Although they conceded that "there seem to be real associations between handedness and some immune disorders, . . . such as those with allergies, asthma, and colitis," they found no associations with arthritis and myasthenia gravis (154–55).

Bryden and McManus insisted that anomalous dominance was a confusing and ambiguous construction, which they asserted was synonymous with left-handedness. Tying specific disorders to left-handedness made no immunological sense. Rather, they suggested that another covariable was probably responsible for the seeming relationship. It is easy, wrote Bryden, "to give some handedness questionnaire to a large sample of captive subjects, score it in an arbitrary way and then to ask about various atopic and immune disorders," but the findings will have no validity. On the other hand, such an exercise may "frequently lead to a publishable paper, but perhaps not a useful one."[47]

Bryden, McManus, and Bulman-Fleming's overall evaluation was direct and harsh: "All things considered, then, we find the evidence to support the Geschwind and Galaburda model lacking and would suggest that psychologists and physicians have more useful things to do than to carry out further assessments of the model."[48]

Annett endorsed the conclusion that further evaluation of the Geschwind hypothesis was an unwelcome distraction.[49] She accused GBG of having presented a confusing and overly complex explanation of her theory (which we discuss in chapter 8). Left-handedness, she asserted, can more easily be explained by chance than by the Geschwind hypothesis.

Annett insisted that Geschwind had hampered serious research on laterality: "The laterality field," she wrote, "has a long history of intriguing speculations based on small trends which typically do not stand up to tough statistical testing." Geschwind and his collaborators assumed "that tenuous links between handedness, sex differences, and developmental language disorders could be used as solid foundations for a complex net of speculative interpretation" (236). Thus, instead of moving research forward, Annett believed "that the GBG model obscured rather than clarified the field, with its blunderbuss approach to theory and method" (236).

As important, the longer-term impact and popular embrace of the GBG model had undermined careful scholarship and research, making it liable to being dismissed as frivolous. It was "many years since I realized that the left brain/right brain bandwagon had brought the laterality field into disrepute with most serious psychologists." Now Annett feared that "what little credibility remained has been swept away by the rush to follow the GBG model." Indeed, she reported that a major neuropsychological journal had decided no longer to consider submissions on handedness, concluding they were generally based on poor scholarship and overreaching theory. This, wrote Annett, was the real Geschwind legacy (240–41).

In Defense of Geschwind

Geschwind had died a decade earlier and could not respond to Annett or McManus. Although Galaburda did not contribute to the 1994
McManus and Bryden volume, he had written a long defense of the
GBG hypothesis a few years earlier in collaboration with two French
colleagues, Michel Habib and Florence Touze.[50] Galaburda reminded
his readers that from the outset the GBG hypothesis was based on
the recognition that both genetics and environment play a mutually
inclusive role. There was, wrote Galaburda and his colleagues, "a complex interaction between biological predisposition, which includes
genetic and (non-behavioral) epigenetic factors, some of which act
in utero, and subsequent learning, which is influenced by the nature
of the task to be accomplished and by social pressures." It is crucial to
recognize, noted Galaburda, that there is a "right-handed hegemony"
in which many activities are culturally structured to fit the fact that
most humans are left-brained. "It is these factors we call 'normal' and
all others 'aberrant'" (122).

Galaburda's defense of his and Geschwind's hypothesis appeared
in a volume of the annual *Advances in Psychology,* edited by psychologist Stanley Coren.[51] Titled *Left-Handedness: Behavioral Implications
and Anomalies,* the collection was a spirited defense of the GBG hypothesis. Coren, rather than Galaburda, who by now had increasing
doubts about the hypothesis,[52] would popularize and defend it (see
page 94 for a photo of Coren).

Ambiguous data supporting the relationship between handedness and immunity, according to Coren, resulted from "different
definitions" of handedness, which produced "different statistical outcomes."[53] Pointing to his and others' research that employed more
rigorous classification standards, Coren found that many subjects fit
a third category, which Coren labeled as "non-right-handed." Including this group, Coren claimed, would provide convincing evidence
in support of the GBG hypothesis (168–69). Studies using sufficient
statistical power, wrote Coren, reported "significant associations between handedness and such immune system related variables as
allergies" (173).

Coren concluded that for the GBG hypothesis to be fairly tested,
the data on both handedness and immunity had to be standardized
(173). But neither Annett nor McManus believed that any further

examination of the Geschwind hypothesis was warranted. And, by the end of the twentieth century, with the renewed emphasis on the role of genes in handedness, the theoretical tide was flowing in their favor. Galaburda, whose own work on dyslexia focused on genetics, no longer defended the GBG hypothesis. "We know things today we couldn't know then," Galaburda told me. "It was good for the time but everyone was wrong."[54]

Both supporters and critics ignored Geschwind and Galaburda's commitment to the individual clinical case and their critique of population studies. They failed to engage Geschwind's insistence on examining all claims in the context of pathophysiological mechanisms. Instead, critics and advocates constructed meta-analyses of subsequent population studies that did not meet Geschwind and Galaburda's reliability criteria.

Geschwind's analyses had threatened to call into question the validity of the genetic models of his critics. The stakes were high. A simpler (earlier) version of the GBG hypothesis that focused on left-handedness was much easier to defend or to dispose of. Rejecting GBG had become the preoccupation of Marian Annett and her countryman, Chris McManus, whose genetic models are the subject of chapter 8.

8

Genetic Models and Selective Advantage

For those who think theories are invented (out of the head of a thinker or perhaps handed down as an inspiration from sources on high) I would point out that this particular development of the [right shift] theory depended on a research process that began 17 years earlier.

MARIAN ANNETT (2002)

The best-understood way in which polymorphisms are balanced is by what is called "heterozygote advantage." Individuals with one copy of each gene . . . are fitter than homozygotes. . . . The classic example is sickle cell anaemia: the person with one sickle cell and one standard gene are fitter overall because they are more resistant to malaria than individuals without the sickle gene, and they do not suffer the complications of having two copies of the sickle gene.

CHRIS MCMANUS (2002)

As a child I assumed that my left-handedness was connected with my mother's, but I did not know how. My father and all my grandparents were, so far as I could tell, right-handed, as was my younger brother. One of my mother's three bothers was left-handed, as were two of my four first cousins on my mother's side. To be left-handers in my extended family one did not necessarily have a left-handed parent or grandparent, but if one did the chance of being left-handed increased. Of course, it is possible, even probable, that either of my mother's or father's parents, born in Europe in the last decade of the nineteenth century, had been retrained to be right-handers. In any case, left-handedness in my family fits the inheritance pattern illustrated in table 1, in which the chance of left-handedness is increased if either or both of a person's parents is left-handed. If both parents are right-handed the percentage of left-handed offspring is approximately 9.5 percent, whereas if both parents are left-handed, the percentage of left-handed offspring increases to 26.1 percent. If the father is left-handed and the mother is right-handed, offspring are 19.5 percent left-handed; this percentage is slightly higher if the mother is left-handed and the father right-handed. In each scenario the percentage of left-handed male offspring is higher than female offspring (the ratio is approximately 5 to 4).[1]

Table 1. Inheritance of Left-Handedness

Handedness of Parents	RH/RH	LH/LH	Father LH Mother RH	Mother LH Father RH
Percentage of Left-Handed Offspring	9.5%	26.1%	19.5%	20%

Source: I. C. McManus, "The Inheritance of Left-Handedness," *Ciba Found Symp* 162 (1991): 257.

This finding is not new—it was the basis for many psychologists' skepticism toward Geschwind's hypothesis. However, as we saw in chapter 6, although there was a widespread belief among researchers that handedness was hereditary, no persuasive explanation had endured the test of time. In large part the popularity of the Geschwind hypothesis rested on its explanatory power, in contrast with genetic models. Nevertheless, research on the heritability of handedness continued. By the early 1980s, simultaneous with the emergence of GBG, there was a sustained effort to develop mathematical models that would reveal the mechanisms for a genetic etiology of handedness. Investigators doing this work shared a belief that the GBG hypothesis not only was wrong but distracted researchers from uncovering the etiology of left-handedness.

With the ascendancy of molecular research in the 1990s, the pendulum moved away from the Geschwind hypothesis toward genetic models. From a genetic perspective left-handedness was assumed to result from either a disruption of the dominant gene for right-handedness or from an inherited, but not necessarily Mendelian, recessive gene. Genetic investigators believed that their models ultimately would guide geneticists to the gene—or genes—for handedness.[2]

The two most prominent model builders were GBG's most persistent critics, British psychologist Marian Annett and psychologist and neurosurgeon Chris McManus. Each assumed that there were

Marian Annett proposed that the prevalence of right-handedness and left-hemisphere speech resulted from a single gene (*RS+*) that impaired the right hemisphere. (Photo courtesy of Marian Annett.)

Chris McManus challenged Annett's RS hypothesis, arguing instead that there was a dominant gene (*D*) for right-handedness and a recessive gene (*C*) for either-handedness. McManus elaborated his thinking about handedness and laterality in professional articles and in his 2002 book, *Right Hand, Left Hand: The Origins of Asymmetry in Brains, Bodies, Atoms, and Cultures.* (Photo courtesy of Chris McManus.)

unidentified genes for left-brain dominance that were reflected in right-hand preference. Each aimed to establish his or her own hypothesis as the most likely explanation for the distribution of handedness. Although these models could not identify specific genes implicated in handedness, they established mechanisms that were consistent with a genetic etiology of handedness.[3] Without plausible hereditary models, it would be difficult for geneticists to identify a candidate gene (or genes) for handedness. Thus a recent discussion of genome-wide association studies (GWAS) of handedness cited only the Annett and McManus models, with modifications, as enabling this research.[4] We begin with Annett's model.

ANNETT'S RIGHT SHIFT THEORY

First proposed in 1979 and elaborated in 1981, Annett's "right shift" (RS+) model rejected both dominant-recessive (Mendelian) and polygenic (multigene) explanations. Instead, Annett initially hypothesized that the predominance of human right-handedness resulted from a

single gene favoring left-hemispheric speech, which she labeled "RS+."
The absence of the *RS+* gene, which Annett labeled as *"RS–,"* did not
code for either side; thus *RS–* could result randomly in either right- or
left-handed preference.[5]

Annett's theory resulted from her clinical work in the 1960s with
children whose epilepsy could be traced to a single location in one
brain hemisphere. Annett found that approximately one-third of these
patients were mixed-handed. Initially, Annett assumed that "lateral
asymmetries in mixed-handers might be due to variable expression
in heterozygotes" for right- and left-handers. But she soon concluded
that it was difficult to test this hypothesis because there was great
variability in the reported estimates of the incidence of left-handedness, with claims running from 1 to 40 percent. Annett decided that
one of her first tasks was to identify reliable data on the extent of left-
handedness. This led her to develop a survey instrument (discussed
in chapter 3) that would, along with the subsequently developed
Edinburgh survey, become the standard for ensuring handedness
classification reliability.[6] Armed with reliable data Annett produced
a series of impressive publications, reporting the results from her
studies of handedness in children with two left-handed parents[7] and
her examination of handedness in twins.[8] Through a process of elimination, Annett became increasingly persuaded that a single gene that
retarded the right hemisphere was responsible for right-handedness.
Its absence or mutation appeared to allow for left-handedness and
ambidexterity and was responsible for a variety of learning disorders,
including schizophrenia and autism.[9]

Annett modified her right shift theory based on her clinical studies of children over the next three decades. The prevalence of right-
handedness and left-hemisphere speech, she concluded, resulted
from a single gene (*RS+*) that impaired the right hemisphere.[10] This
suspect gene was responsible for the normal pattern of hemispheric
specialization and only incidentally for right-handedness. Thus, for
Annett, strong right-handedness was a proxy for left-hemispheric
dominance.[11]

Annett assumed that those lacking the *R+* gene were generally left-
handed or non-right-handed. They also were at risk for developing
learning disorders and held the potential to develop special talents. As
evidence of the latter she pointed to a higher prevalence of non-right-
handed mathematicians, which she connected to the absence of the

RS+ genotype.[12] As she wrote, "Evidence for heterozygote advantage has been found in studies of educational achievement, especially in reading. Poor readers are more prevalent at both extremes of the laterality distribution" (427).

In her revisions, which appeared in her comprehensive 2002 book *Handedness and Brain Asymmetry: The Right Shift Theory,* Annett postulated three possible alleles. The first, *RS−,* was an "ancestral primate allele," which is neutral for laterality. The second, the human allele (*RS+*), impairs the development of the right hemisphere and weakens the left hand, thereby strengthening the right hand. Finally, the *RS+*a, or "agnosic," gene carries instructions to impair either hemisphere at random. This revised model led Annett to develop a hypothesis that human hemispheric asymmetry results from "*right hemisphere disadvantage* rather than left hemisphere advantage"—although the exact mechanisms are unknown. Inheritance of a combination of the *RS+* and the *RS+*a, Annett speculated, can sometimes result in impairment of both hemispheres, resulting in schizophrenia, while inheritance of two *R+*a alleles can result in autism and associated learning disabilities.[13] (We examine the connections Annett made between her RS hypothesis and schizophrenia and autism in chapter 11.)

Environment	- main cause of lateral asymmetries, universal in primates including humans - asymmetries arise from random accidents of embryological growth - these lead to a Gaussian distribution of R-L differences
Genes at RS locus **RS−** **(rs minus)**	- ancestral primate allele(s), neutral for laterality
RS+ **(rs plus)**	- a human species allele - it carries an instruction to "impair the growth of the *right* hemisphere" - left handed weakness shifts the R-L distribution toward dextrality
RS+ a **(rs agnostic)**	- a mutation of *RS+* - it gives an instruction to "impair the growth of *one* hemisphere" - the left *or* the right hemisphere is impaired at random

An illustration of Annett's right shift hypothesis.

McManus's Dextral/Chance Model

McManus and Annett have carried on a vigorous debate in academic journals for the past three decades.[14] In McManus's dextral/chance (DC) model, which challenged Annett's RS hypothesis, McManus argued that there was a dominant gene (*D*) for right-handedness and a recessive gene (*C*) for either-handedness. Thus, depending on one's parents' genes, there are three possible combinations, *DD*, *DC*, *CC*, and three possible offspring results: *DD* = results in right-handedness, *CC* = 50 percent chance of left-handedness, and *DC* = 25 percent chance of left-handedness.[15]

Like Annett's model, the McManus model also developed over time. Both models ultimately attempted to account for the sex-related differences in left-handed prevalence.[16] Also, like Annett, McManus provided an explanation for the putative prevalence of learning disorders associated with left-handedness. McManus also offered a likely selective advantage for the persistence of left-handedness.

McManus elaborated his thinking about handedness and left-hemisphere laterality in numerous professional articles and in his 2002 book, *Right Hand, Left Hand: The Origins of Asymmetry in Brains, Bodies, Atoms, and Cultures.*[17] Bringing much of McManus's thinking together, the book is a model of argument and erudition. It is essential reading for those interested in the origins and impact of handedness, and in scholarship on the subject. Here McManus lays out the logic for the selective advance of his chance gene hypothesis, which, he admits, is purely speculative although scientifically plausible (228).

The greatest advantage of *DC* genotype (heterozygote), according to McManus, is that it confers randomness on brain organization. This effect is not restricted to language and manual dexterity but includes other asymmetries such as reading, writing, visual-spatial processing, and emotions. Small amounts of randomness, writes McManus, are beneficial to complex systems. Thus, while *CC* type is disadvantageous, it is necessary for the existence of *DC* type, which is potentially advantageous to its recipients.

A bimodal effect of the *DC* type, according to McManus, can account for both the highly gifted and those with learning disabilities. As he explains, the upside of randomness is that some modular components of language probably work better close to one another, that is, in the same hemisphere. Thus, McManus's randomness can account for improved brain performance, resulting in gifted individuals. This effect is not limited to the left-handed, but rather it includes the entire *DC* genotype, which results in 75 percent right-handers. Too much hemispheric randomness, however, is not beneficial and may result in a variety of learning disabilities, including dyslexias, stuttering, autisms, and schizophrenias.[18]

A SELECTIVE ADVANTAGE?

Although the Annett and the McManus models are each persuasive, they stand in contradiction to each other. Moreover, their models' critics point out that neither model has fully accounted for anomalies, particularly the sex difference and epidemiological variances. But a more important question is unanswered by both models: why are there any left-handers and why are there always many fewer of them? Both Annett and McManus see the answer in the heterozygote phenotype. Based on extensive testing of schoolchildren of all ages and grades, Annett finds that the most successful students were less strongly right-handed than their peers. Examining their pedigrees, she reports that these successful students were more likely to be heterozygotes.[19]

McManus takes a more historical approach, opening the door to the role of environmental pressures to account for the changing prevalence of left-handers. McManus endorsed the claim that the numbers of left-handers precipitously declined in the nineteenth-century United States.[20] He attributed this to the stigma toward left-handers as reproductive partners, reducing the *CC* genotype and resulting in a steep decline in the left-handed population.[21] McManus and colleagues differentiate those societies where left-handers were merely forced to adopt right-handedness for tasks such as writing and eating (where their genes would nevertheless continue to be passed down) with those societies in which left-handers would be viewed as less desirable marriage partners and would therefore produce fewer offspring. By incorporating an environmental mechanism (discrimination) for the prevalence of left-handers, McManus allows for the possibility that the prevalence of left-handers depends on genes *and* environment. In particular he has been intrigued by a recent interpretation of the so-called fighting hypothesis.[22]

As McManus and his colleagues recently noted, despite being periodically dismissed, the fighting hypothesis has had remarkable resilience.[23] Originally proposed by the nineteenth-century Scottish historian Thomas Carlyle and simultaneously by his colleague Philip Henry Pye-Smith, then vice chancellor of the University of London, the theory assumed that most warriors held their swords or other weapons in their right hands, protecting their left side (often with a shield). Thus their heart, situated on their left side, was protected, whereas left-handers would expose their hearts to their right-handed combatants, resulting in a selective advantage to right-handed warriors.[24]

Both Gould and Parson (whom we discussed in chapter 6) had invoked this theory. Most subsequent commentators found the shield/sword hypothesis to be unconvincing. Psychologist Stanley Coren called the hypothesis ahistorical because before the shield was invented, rocks and sticks served as weapons. Neither right-handers nor left-handers are offered any special protection during stone throwing.[25]

Despite these criticisms, French evolutionary biologists Charlotte Faurie and Michel Raymond have recently suggested that an elaboration of the fighting hypothesis explains why there is any left-handedness at all and why left-handers remain a minority at the population level. They argue for a negative frequency–dependent advantage, in which the advantage of left-handedness decreases as the number of left-handers increases.[26] According to Faurie and Raymond, this situation comes about from the surprise effect that left-handers have in combat and in sports. That advantage is retained as long as the population of left-handers remains small. However, as the number of left-handers increases, they lose their selective advantage. Faurie and Raymond point to interactive sports where left-handers are overrepresented, including wrestling, fencing, and baseball, as proxies for traditional combat. (They admit that a further test of their hypothesis would be to determine the winning rates of left-handers in sports.)

Faurie and Raymond also examine fighting in traditional populations, such as the Eipo in Papua, New Guinea. They find that both violence and left-handedness were more prevalent in Eipo culture prior to contact with industrial societies. They assert that colonization resulted in a rapid decrease of traditional violent social practices and thus of "frequency-dependent selection associated with handedness" (110).

Reviewing Faurie and Raymond's hypothesis, McManus and colleagues find it "an elegant evolutionary hypothesis." They particularly are intrigued by Faurie and Raymond's explanation that left-handers, because of their smaller numbers, provide surprise and thus a greater chance of victory in combat and sports. While they concede that "evidence for the fighting hypothesis is not particularly strong," they nevertheless admit "there is little evidence to reject it either." Thus, they conclude that while the fighting hypothesis "remains an intuitively plausible explanation for the persistent left-hand preference in the population," validating it would require much more data, including examinations of the fighting behavior of nonhuman primates.[27]

The attraction of the Faurie and Raymond hypothesis for McManus is that it provides a possible solution to the issue of selective advantage. Annett does not discuss the negative frequency–dependent advantage hypothesis, but it is consistent with her RS theory. That is, if an ancient trait or specific gene or genes, such as Annett's $RS-$ or McManus's ancient C gene, were responsible for left-handedness, left-handedness would persist only if it conferred a heterozygous advantage. Despite its admittedly speculative basis, the fighting hypothesis remains attractive because it combines biological and evolutionary factors to explain the persistence of left-handedness. This interpretation of the fighting hypothesis provides an illustration of how environmental pressures can account for the anomalies that exist in genetic models, like those of Annett and McManus.

The revised fighting hypothesis, with its environmental features, was not extended to include the Geschwind hypothesis. Nevertheless, while the interest in the Geschwind hypothesis is more historical than applied, the issues that GBG raised—particularly the relationship between handedness, learning disabilities, and creativity—continue to engage researchers.

Looking back over almost a century of research, Geschwind and Galaburda had found no satisfactory explanation for left-hemispheric dominance in humans. Among their contemporaries' work they preferred the explanations of psychologists Michael Corballis (New Zealand) and Michael Morgan (Britain), that an early insult damaged or restricted normal development of the left hemisphere, enabling more rapid growth of the right hemisphere.[28] For his part Corballis was intrigued, but not persuaded, by the Geschwind and Galaburda hypothesis.[29]

Although Annett uncompromisingly rejected the GBG hypothesis, Corballis initially believed that the two types of explanations were complementary, in that GBG provided a putative mechanism (hyper-transmission of testosterone) for disabling normal left-hemispheric dominance. Skeptical of Annett's claim that her hypothetical *RS*– allele accounted for learning deficits and mental disorders,[30] Corballis nevertheless believed that Annett's *RS*– model best fit the frequency and distribution of handedness and left-hemisphere dominance for language.[31]

Though I have often relied on Michael Corballis's work in this book, I have not examined his specific contributions. In his classic 1991 book, *The Lopsided Ape: Evolution of the Generative Mind*, Corballis argued that handedness and hemispheric laterality were unique to humans.[32] A quarter century later Corballis renounced his earlier hypothesis and concluded that neither handedness nor hemispheric laterality were restricted to humans.[33] Following Corballis's journey as he arrived at this position provides a window to explain why handedness studies have moved from environmental concerns, including the GBG hypothesis, to an almost exclusive genetic paradigm, as evidenced by the ultimately positive reaction to both the Annett and McManus models. Although researchers continue to concede a role for environmental influences, the research itself focuses on the hypotheses and methodologies of genetics. Thus, the earlier suggestions by British psychologists Margaret Clark and G. C. Ashton that there may be multiple routes to left-handedness are rarely examined.[34] Corballis's change of mind, which we turn to in chapter 9, illustrates why this is so.

9

Uniquely Human?

Although there may be weak precursors of handedness as well as left-hemispheric specialization for vocalizations in other species, the combination of the two asymmetries seems to be unique with humans.

MICHAEL CORBALLIS (1991)

One myth that persists even in some scientific circles is that asymmetry is uniquely human.... [However,] a right hemisphere dominance for emotion seems to be present in all primates so far investigated, suggesting an evolutionary continuity going back at least 30 to 40 million years.

MICHAEL CORBALLIS (2014)

A re right-hand dominance and hemispheric asymmetry uniquely human? And does it matter? I turn to the work of the influential and widely respected New Zealand psychologist Michael Corballis, initially an ardent advocate of uniqueness[1] but recently converted to the opposite view.[2] Corballis's intellectual journey is emblematic of controversies among researchers over the past four decades, and examining his conversion reveals what is at stake from both a scientific and a philosophical perspective.

Corballis's earlier work was powerful and persuasive. His revisions reveal much about how changing scientific paradigms have shaped hypotheses about human laterality and language. As I elaborate below, a paradigm shift elevated the role of genetics and mirror neuron theory while modifying the definition of evolutionary biology. The debate exposes subterranean issues related to human choice and behavior or what traditionally was labeled free will and predestination but today is played out as cultural framing versus biological reductionism. Indeed, Corballis acknowledges this subtext in his recent article "What's Left in Language?" in which he cites the argument of the eighteenth-century philosophers Étienne Bonnot de Condillac and Giambattista Vico, both of whom believed that human language had evolved from primate and "primitive" human gesture.[3] Whether or not this is so is central to Corballis's revision of Corballis.

UNIQUELY HUMAN

Corballis had hypothesized that bipedalism liberated the two hands from locomotion, freeing humans to use their hands for specialized tasks, such as tool making and tool use.[4] Task specialization encouraged hemispheric laterality. Corballis insisted this was a uniquely hominid phenomenon that enabled development of language, which Corballis differentiated from animal communication.[5] Conceding that many animals communicated via gesture and sound, Corballis contended that only humans did so through a sophisticated symbolic system we call language.

Unlike other primates, Corballis pointed out, only humans demonstrated significant hemispheric laterality, in which approximately 90 percent were right-handed in motor functions while 95 percent were left-brained language dominant. Since human language

Michael Corballis, the New Zealand psychologist whose many books and publications tied bipedalism to the development of uniquely human lateralized language. Lately Corballis has revised this claim. (Photo courtesy of Michael Corballis.)

overwhelmingly resided in the dominant left hemisphere, Corballis was persuaded that language itself was a consequence of laterality. Because there seemed to be only a small anatomical difference between *Homo sapiens* and nonhuman primates, Corballis assumed that human motor and linguistic asymmetry developed because of environmental and evolutionary pressures. Thus, human bipedalism created the conditions that made language possible but not inevitable. This view seems compatible with the "exaptation" hypothesis of paleontologists Stephen Jay Gould and Ian Tattersall. Gould defined an exaptation as "a feature, now useful to an organism, that did not arise as an adaptation for its present role, but was subsequently coopted for its current function."[6] Tattersall distinguished exaptations from adaptations similarly: "*Adaptations*," wrote Tattersall, are "features that were fixed in the context in which they are now employed," while exaptations are "features that originally arose in one context but were later coopted for use in another."[7] Although Corballis did not embrace the term *exaptation,* his hypothesis fit Gould's definition: "The most parsimonious explanation for so-called right-hemisphere specialization," wrote Corballis in 1980, "is simply that it is gained by default. The left hemisphere, having assumed the dominant role in purposive actions, may have forfeited some of its capacity for functions that would otherwise be bilaterally represented."[8]

Exaptation could also account for the uniqueness of human laterality. In a series of articles and books, beginning in the early 1980s, Corballis reported traces of laterality as early as two to three million years ago, in *Homo habilis.*[9] Thus, while Corballis conceded that "nonhuman animals exhibited some asymmetries," he described these as "weak precursors of human laterality in our shared ancestors." For Corballis writing in 1989, "the pattern observed in humans did not emerge until the hominid line split from the other primates." Among primates, "only humans," wrote Corballis, "show the marked population bias toward right-handedness" and asymmetry.[10]

The function of genes in the etiology of handedness, he wrote, remained "elusive."[11] Conceding that genes played a role in this process by making human linguistic laterality and right-handed dominance possible, Corballis insisted that they did not determine the shape or use to which laterality and hand dominance would be put. Noting that others had reported that "handedness and cerebral asymmetry are not unique to humans," Corballis asserted that "activities that are uniquely human, such as language, or that are more highly developed in humans than in other species, such as manual skill (dexterity), may well have exploited asymmetry in ways not evident in non-human species."[12] The emergence of human language was randomly determined—it was one possibility, influenced by a genetic blueprint but not predetermined by it. Although Corballis did not endorse the Geschwind and Galaburda testosterone hypothesis, he did not reject it either, leaving the door open for a revised trauma hypothesis.[13]

Corballis found much of value in both Annett's single-gene hypothesis and McManus's two-allele models.[14] He applauded both models for offering plausible explanations for the distribution and persistence of left-handedness, even if neither answered the specific criteria Corballis had outlined.

He proposed an alternative "evolutionary scenario," and for the sake of simplicity he adopted McManus's allele identification of (D = dextral/right and C = 50% chance of right or left). "If the D allele emerged early in hominids, eventually spreading to 90 percent of the population," wrote Corballis, it should have "conferred an advantage on those inheriting it." If so, it would have replaced all the alternative genes. The persistence of the heterozygotic (CD), however, suggests it must confer some advantage. "That is," wrote Corballis, "CD individuals must have greater fitness than either CC or DD homozygotes, 'fitness' referring to the relative number of offspring contributed by an individual to the next generation" (718). (This, of course, was the same claim that McManus made. See chapter 8.)[15]

Corballis conceded that his case for a heterozygous advantage rested on probability not empirical evidence. He was persuaded that while the transition to vocal language was most likely assisted but not determined by a genetic mutation that ensured vocal and manual gestures, it would be localized on the same side.

Over the years Corballis modified his analyses, leaving room for the possibility of more transspecies origins of laterality.[16] A conference of laterality researchers held in Germany in 2011 focused on the issues that framed Corballis's writings: were handedness and linguistic laterality uniquely human, the unintended result of environmental pressures? This conference can help illustrate the context in which Corballis changed his mind.

The Delmenhorst Conference

Much recent handedness scholarship debates human uniqueness and the role of genes, as illustrated in the proceedings of an October 2011 symposium on handedness sponsored by the Hanse-Wissenschaftskolleg Institute for Advanced Study, in Delmenhorst, Germany. The meeting was organized by biological anthropologists William C. McGrew of Cambridge University, Wulf Schiefenhövel of the Max Planck Institute, and Linda Marchant of the University of Miami, Ohio. The revised presentations were published in the *Annals of the New York Academy of Sciences* in 2013. The Delmenhorst symposium symbolically signals a final (for now) defense of the human exceptionalism approach; after this, research on handedness would turn toward genetic foci that included nonhuman animals.

Six of the papers examined nonhuman primates for clues for precursors that evolved into human handedness and linguistic laterality. Others focused on whether human language and motor laterality were "a by-product of other selective forces, such as those shaping language or tool use"—in other words, an exaptation.[17]

The contributors affirmed previous claims that all human populations included a minority of left-handers whose proportion of the population varied according to the definitions and instruments employed to identify handedness. McGrew and colleagues argued that handedness required cultural and environmental pressures but conceded a possible role for polymorphisms. They concluded that a comprehensive examination of "traditional, preliterate societies" was needed in order to eliminate the cultural biases of contemporary data that often were tainted by the forced switching of left-handers, schooling, and other industrial requirements.[18]

The presentations examined the alleged motor asymmetry and lateralization in great apes that varied depending on context and methodology.[19] American primatologist William Hopkins reported that he continued to observe hand preference among chimpanzees at a population level but conceded that "the expression of handedness is less robust compared to humans."[20] Hopkins's views were supported by University of London anthropologists Jeroen B. Smaers and colleagues, who pointed to increasing evidence of motor lateralization in vertebrate brains. Reviewing primate evolution over the past 46 million years, they found an increasing role of the frontocerebellar areas in apes beginning with their separation from monkeys approximately 30 million years ago (~30mya). They also pointed to increasing involvement of the prefrontal cortex over frontal motor cortex in both the Homo-Pan (~10mya) and the ancestral human lineage (~6mya).[21]

In contrast, a joint Cambridge University archeological and anthropological team led by Jay T. Stock reported consistent right-handed asymmetry throughout human history not found in chimps. They pointed to "mechanical differences between early human hunter-gathers and *P. troglodytes,* which revealed that while the chimpanzees were left-lateralized in the morphology of the humerus and right-lateralized in the second metacarpal ... all human populations are predominantly right-biased in the morphology of these bones." The team attributed these differences to a human response to technological developments.[22]

Placing these findings in a wider context, anthropologists Marchant and McGrew noted that whether human handedness was unique compared to chimpanzees and bonobos depended on the definition of handedness. They suggested that "handedness (congruence across subjects and across tasks) should be distinguished from hand preference (within subject and task), manual specialization (within subject, across tasks), and task specialization (across subjects, within task)." Making these distinctions revealed that although there was much evidence "for task specialization in chimpanzees," there is "no conclusive evidence of handedness in the strictest sense. Thus, human handedness seems to be unique among living hominoids."[23]

Austrian American biologist W. Tecumseh Fitch and American anthropologist Stephanie N. Braccini concluded, as Corballis had earlier, that compared with humans "manual specialization in nonhuman primates is relatively weak. A right-bias in chimpanzees may exist, but is so weak that many studies using simple tasks fail to reveal it."[24]

Nevertheless, the advocates of chimp laterality as a precursor to human asymmetry remain persistent. Hopkins and his colleagues have continued to refine their arguments. Writing in 2015, they argued that "contrary to many historical views, recent evidence suggests that species-level behavioral and brain asymmetries are evident in nonhuman species."[25] This finding, argues psychologist Clare Porac, "support[s] theories that emphasize the importance of tool use rather than gestural communications as the evolutionary adaptation that gave rise to right-handedness in humans."[26]

The presentations on human handedness reflected a division between those who attributed it to culture versus those who emphasized
genes. French psychologist Jacqueline Fagard found right-handed
preference in all human societies and cultures, which she attributed
to constant reinforcement beginning in the uterine environment continuing and intensifying in the family and wider culture.[27] In contrast, British psychiatrists Thomas H. Priddle and Timothy J. Crow
endorsed Annett's theory of a right shift gene and speculated on the
identity of the likely genetic candidates that code for asymmetry.[28]
Not surprisingly, Chris McManus and colleagues found his own model
to provide the most persuasive mechanism for the persistence of left-
handedness. Handedness, they wrote, was determined by a combination of genes that coded for "height, primary ciliary dyskinesia, and
intelligence." As they wrote, "his theory suggests that handedness
inheritance can be explained by a multilocus variant of the McManus
DC model, classical effects on family and twins being barely distin-
guishable from the single locus model."[29]

Finally, as discussed in chapter 8, the French evolutionary biologists Charlotte Faurie and Michel Raymond, committed to exploring
ways to connect environmental and inherited causes of left-handedness, resurrected the "fighting hypothesis." Their earlier investigations
pointed to the human uniqueness of left-handedness, tracing it back
to the Upper Pleistocene (90,000 to 100,000 years ago).[30] Here they
hypothesized a left-hander "frequency-dependent advantage" so that
left-handers initially have an advantage in a fight that decreases as the
number of left-handers increases.[31]

Faurie and colleagues built on their 2011 study that examined testosterone in the saliva of 64 French university students. They found, in
seeming partial support for the Geschwind hypothesis, "a significantly
higher average testosterone concentration in left-handers than in
right-handers, consistent with frequencies of fights." This suggested
to Faurie that "these behavioral and hormonal differences may be acquired throughout life due to previous experiences in a social context
and may favor the persistence of left-handers in humans."[32] Others at
the meeting were skeptical about the Faurie hypothesis but not necessarily with the assertion of the human uniqueness of left-handedness.
Dutch behavioral biologist Ton G. G. Groothuis and colleagues pointed
out a logical flaw in the argument: "The fighting hypothesis," they asserted, "assumes that left-handers, being in the minority because of

health issues, are still maintained in the population since they would have a greater chance of winning in fights than right-handers due to a surprise effect."[33]

The validity of the fighting hypothesis rests on the assumption that increases in testosterone levels contribute to the incidence of left-handedness. Usually, one might assume that sustained increases in testosterone take place earlier, either in utero or early childhood. From this perspective, the impact of fighting results in a selective advantage for left-handedness because it ensures and exacerbates the inheritance of a putative set of left-handed genes by the fighters' offspring. If higher levels of testosterone play an epigenetic role in the etiology of left-handedness, then any mechanism that permits such a scenario, including uterine trauma, could be implicated in the etiology of handedness. This is, minus the fighting hypothesis, the claim made by Geschwind and his colleagues in the early 1980s (see chapter 7).

Readers of the papers from the Delmenhorst symposium would learn that there was no consensus about the uniqueness of human handedness and hemispheric laterality. Studies produced since the publication of the symposium papers, however, seem unanimous in assuming a genetic substrate that will be revealed by focusing on both putative genes for asymmetry and closer examination of animal preferences. Thus, the Brandler studies, discussed in chapter 1, identified the supposed gene for cilia asymmetry, whose absence they hypothesized manifested itself in human left-handedness.[34] Meanwhile, Hopkins and his colleagues at Emory University's Yerkes National Primate Research Center in Atlanta, Georgia, continued to report more evidence of population-wide hand preference among nonhuman primates.[35] They especially pointed to mirror neuron theory, which I discuss below, as confirming the origin of speech in gesture found in both humans and great apes.[36] Finally, the kangaroo study, discussed in chapter 1, interpreted left-forearm preference in bipedal marsupials as similar to the cause of handed asymmetry in humans.[37]

Given the wider assumption among researchers that the key to human behaviors would be uncovered in the human genome, there has been diminished support for identifying environmental causes of handedness and linguistic laterality. The action was now in the human genome, and researchers, including Corballis, who wanted to remain relevant had to consider the role of the human genome and animal precursors to human laterality more sympathetically. They also would have to engage the tools and hypotheses of the genetic paradigm shift, the Genome-Wide Association Study, brain scanning, and mirror neuron theory. The exaptation hypothesis has almost disappeared from the handedness discourse.

BACK TO CORBALLIS AND THE ROLE OF GESTURE

Although Corballis did not participate in the Delmenhorst Conference, the issues debated were central to his continuing examination of human uniqueness. This is evident in his publications in 2014 and 2015, the latter of which appeared in the same journal, *Annals of the New York Academy of Sciences*, as the published symposium. As throughout his career when new evidence emerged, Corballis did not resist its implication even when it contradicted his work. Contrary to his earlier conclusions, Corballis now has become persuaded that there are other species with similar asymmetries to humans. While he continued to believe that "complex" linguistic and manual skills were uniquely human and are found in the asymmetric brain, he now reported that a number of asymmetries identified in other species appear to be "precursors" of human behaviors, while others may be similar in humans and nonhumans. Corballis suggested that these shared "asymmetries are best understood in the context of the overriding bilateral symmetry of the brain and body, and a trade-off between the relative advantages and disadvantages of symmetry and asymmetry." This way of thinking provides a framework for constructing the genetics of handedness. Recent genetic models, writes Corballis, have postulated a single gene with two alleles, one favoring right-handedness and the other either left- or non-right-handedness. But Corballis remains skeptical that this gene has been identified or located. Instead, he suggests that a number of genes are probably involved "or that the gene may be monomorphic with variations due to environmental or epigenetic influences." Thus, Corballis concludes, handedness may best be understood as resulting from a combination of genetic and environmental pressures, making it "more profitable to examine the degree rather than the direction of asymmetry."[38]

From this perspective Corballis increasingly was persuaded by the decades of research on chimpanzees that implicated a similar laterality to that found in humans. In an article published in *PLOS Biology* in

2014, Corballis presented an overview of his latest thinking on handedness.[39] "Handedness and brain asymmetry," wrote Corballis, "are widely regarded as unique to humans, and associated with complementary functions such as a left-brain specialization for language and logic and a right-brain specialization for creativity and intuition. In fact, asymmetries are widespread among animals, and support the gradual evolution of asymmetrical functions such as language and tool use." Thus, he concluded, unlike in his earlier assumptions, that "handedness and brain asymmetry are inborn and under partial genetic control," even if "the gene or genes responsible are not well established."[40]

The authors of the 2015 kangaroo study pointed to left-forelimb preference among tree kangaroos as the connection of bipedalism and the development of laterality.[41] But having identified these features in marsupials, its authors came to the opposite conclusion, which Corballis had reached earlier. That is, that neither bipedalism nor laterality was restricted to humans. For his part Corballis finds confirmation of his own work in the kangaroo study's suggestion of a connection between marsupial and human handedness. Corballis also endorses the claim by the kangaroo study's hypothesis "that specialization of forelimb functions in red-necked wallabies have been shaped under the pressure of ecological factors."[42] In fact, the kangaroo study might as easily serve as evidence for the multifactorial, or even randomly determined, etiology of laterality and hand preference, especially because in these marsupials the left hand was dominant—the opposite of handed preference in humans.

But *why* was human lateralization, unlike that in these kangaroos, right-handed while left-brained for language? The most persuasive answer came from Corballis writing 35 years ago, when he concluded that in order to gain left-hemispheric linguistic domination, the left hemisphere forfeited some of its otherwise bilateral functions, such as spatial manipulation, to the right hemisphere.[43]

Given the unresolved nature of the gene-environment debate, why had Corballis abandoned his insistence that motor and linguistic laterality were uniquely human for a more instrumental transspecies evolutionary genetics? Corballis explained that his change of mind was based on recent research findings, wider availability of new imaging technologies, mirror neuron theory, and the reinterpretation of the evidence he had earlier relied upon to demonstrate human exceptionalism. "The trend toward complexity and lateralization," Corballis was now persuaded, "was probably accelerated in hominids when bipedalism freed the hands for more complex manufacture and tool use, and more expressive communication."[44]

He emphasized the role of gesture in the development of speech. There was language prior to speech, wrote Corballis, built on a combination of facial and vocal gestural communication.[45] Corballis found support for his revised view of the role of gesture in mirror neuron theory, hypothesized by Italian researchers in the early 1990s, which posits that when an animal or person observes another perform an action, a mirror neuron is allegedly activated in the observer causing it to mimic the behavior and act as if the behavior were of its own initiation. This hypothesis was based on observations that areas of premotor cortex of macaque monkeys were activated during hand movements resulting in imitative behaviors.[46] Over the following decades mirror neuron theory became attached to humans and hypothesized to be the mechanism that led to empathy and its absence.[47] Mirror neuron theory, wrote Corballis, "suggests that language evolved from a system largely devoted to manual grasping in primates, and extended to manual gesturing, pantomime, and ultimately to human speech."[48] Joining with University of Auckland colleagues in an fMRI study of left- and right-handers performing language tasks, the Corballis team reported a left-cerebral hemispheric bias for language, gesture, and handedness. This fits an evolutionary perspective "in which the primate mirror neuron system (MNS) became increasingly lateralized and later fissioned onto subsystems." The first of these was for language, while the second enabled the observation and execution of manual actions. This in turn was subdivided into one regulating hand preference and another that did not, enabling the human "tripartite system of language, handedness, and praxis."[49]

Although the mirror neuron thesis has attracted wide and enthusiastic endorsement, especially among those involved in studying nonhuman animal behavior (empathy, for instance),[50] the connection between gesture and the development of human language lately has been disputed.[51] Based on her close reading of the experiments that undergird mirror neuron theory, historian of neuroscience Katja Guenther exposes the complexity of theorists' claims. Unfortunately, she writes, many others have applied simpler or more restricted versions of mirror neuron theory. She is particularly skeptical of the conflation of the theory with empathy. The "study of empathy has certainly been the most prominent of the research directions resulting from the discovery of mirror neurons, [but] it is only one strand of a larger paradigm."[52]

Psychologist Karen Emmorey writes that if "the mirror system is key to the evolution and emergence of language in humans, it is puzzling that the mirror system plays such a small role in language processing for both auditory-vocal and visual-manual languages."[53] In fact, she finds "little evidence that sign-related mirror neuron populations (if they exist) play a critical role in the perception and comprehension of sign language" (207). Pointing to psychologist Gregory Hickok and colleagues' findings, Emmorey notes that "deaf patients with damage to Broca's area exhibit sign articulation deficits, but sign perception and comprehension are intact" (206).[54] From this perspective it seems clear that the mirror neuron system does not appear to underlie language perception for either spoken or signed language" (208). Finally, she notes that in mirror neuron theory "the remnants of the gestural origins of language (i.e., pantomimes and modern protosigns) should co-occur with speech, but they do not" (208).

The validity of mirror neuron theory in the development of speech aside, it is difficult to identify substantial new evidence that undermines Corballis's initial hypothesis. That is, compared with humans the hemispheric and motor laterality identified in our closest primate relatives is weak. Moreover, despite left-paw preference in one type of bipedal kangaroo, the much greater opportunities afforded by human bipedalism remain persuasive. What had changed in the past 35 years is the nature of what constitutes evidence and the emergence of new methodologies and analytic tools. Because of the increased focus on the genetic origins of human behavior, the development of Genome-Wide Association Study to search for errant genes is now economical. This development has come simultaneously with huge increases in funding to support genetic research that has encouraged a shift from exaptations to reductionism. I am not implying that Corballis's change of mind was driven by funding sources; rather, he reevaluated the role of genetics in light of the availability of new technologies, including scanning technologies, to elicit and interpret genetic data. These data are compatible with his view of the mirror neuron system. Both methodologies and what constitutes evidence are different today from what they were in the 1980s and 1990s. From this perspective Corballis appears to be reading his own earlier conclusions in light of a new discourse that implicates genetic continuum rather than evolutionary discontinuity.[55]

As this discussion of Corballis and the emerging genetic paradigm indicates, the role and extent of nature and nurture in the etiology of handedness is far from resolved. But the triumph of the reductionism, especially when it comes to funding, has ensured that current and future research into handedness will, as with so many behavioral investigations, focus on genes.

The same can be said about the putative but persistent connection between left-handedness and pathological conditions. Corballis hypothesized that "departures from the 'norm' of right-handedness and left-brain language dominance," especially an absence of asymmetries, played a role in the onset of learning disabilities and mental disorders, or the other way around—that is these disorders led to hemispheric symmetries.[56] Based on his evolving views Corballis has called for a reexamination of Orton's (now rejected) claims in the 1930s that weak asymmetry was connected to the etiology of stuttering and dyslexias.[57] Though he didn't point it out, Corballis's view that learning disabilities and mental disorders are connected to anomalous hemispheric dominance is a restatement of the final Geschwind and Galaburda hypothesis but with a focus on genetics rather than environment.[58]

As laid out by psychologist Murray Schwartz a quarter century ago, two contradictory positions continue: "The hard pathological position" is "that *all* left-handedness is the result of cerebral insult. The soft pathological position, and the more tenable," argues that there are several routes to left-handedness with approximately half fitting a pathological (i.e., brain-damaged) diagnosis. For Schwartz, most left-handers exist because of a combination of "normal genetic variability" augmented by prenatal environmental pressures.[59]

The assumptions and methodologies that have connected left-handedness to abnormality, such as schizophrenia and learning disabilities, have also been used to associate homosexuality with left-handedness. In chapter 10 we examine the construction of this claim, which provides a powerful reminder of both the strengths and limits of using the left hand as a proxy for conditions and behaviors.

10

A GAY HAND?

Several homosexuals have written to us suggesting that there is a high rate of left-handedness in this population, but no study of this claim has yet been reported. A high rate of non-right-handedness in this population may seem at first to be difficult to explain in light of some animal experiments but recent studies have shown that, in rats, stress in mid-pregnancy causes the male off-spring to have permanently low free testosterone levels and homosexual behavior.

GESCHWIND AND GALABURDA (1987)

In my experience the main argument [is] . . . that left-handers are "weird," as in deviating in various ways from what is considered normal. As you know, conformity to the "normal" was still very very big in the late 50s–early 60s. "Flaky" is another popular anti-left-handed stereotype. Homosexual is another. You didn't want your son to be any of these things in the 1950s and 60s.

ROBERT WILLIAMS (JULY 18, 2016)

L eft-Handers Are More Likely to Be Gay" headlined a story in the November 2000 issue of *Psychology Today*, which reported that researchers at the University of Toronto's Centre for Addiction and Mental Health had discovered "that homosexuals are more likely to be lefties than heterosexuals." In an interview for the story, one of the study's authors, psychologist and sexologist Kenneth Zucker, explained that, based on analysis of data in "20 different studies over the past 50 years," they found "that lesbians have a 91 percent greater chance of being left-handed or ambidextrous than straight women, while gay men are 34 percent more likely than straight men to not be right-handed." Although their research provided "empirical evidence that links homosexuality to left-handedness," said Zucker, it didn't explain why this "relationship exists."[1] The study, published earlier that year in the academic journal *Psychological Bulletin*, was more circumspect than *Psychology Today*, concluding that homosexuals were more likely to be left-handed and non-right-handed and "that sexual orientation in *some* men and women has an early neurodevelopmental basis."[2] Before returning to the Toronto study it is useful to review the putative association between left-handedness and homosexuality from a historical perspective.

HISTORICAL PERSPECTIVE

The pejorative meanings attached to the word *left* in all human languages also have been applied to homosexuals: perverted, immoral, abnormal, broken, wrong path, gauche, evil, and devious. (Prejudice and discrimination against left-handers is discussed in chapter 4.) In the late nineteenth century, criminologist Cesare Lombroso, like many of his contemporaries, labeled homosexuality as "social inversion," placing it with criminal insanity, homicidal monomania, kleptomania, nymphomania, satyriasis (excessive sexual desires among men), and mental illness.[3] Given Lombroso's claim that left-handers were more likely to be mentally defective and criminal, it would have been consistent for him to make an explicit connection between homosexuality and left-handedness. He did not do so.[4]

Others were less restrained. In his long correspondence with Sigmund Freud at the end of the nineteenth century, Wilhelm Fliess insisted that left-handers were either overt or repressed homosexuals. "Where left-handedness is present," wrote Fliess, "the character pertaining to the opposite sex seems more pronounced. Since degeneracy consists in a displacement of the male and female qualities, we can understand why so many left-handed people are involved in prostitution, and criminal activities."[5]

Less judgmentally, Geschwind and Galaburda had reported that mothers exposed to uterine stress gave birth to males who had permanently low testosterone levels and homosexual behaviors accompanied by high rates of non-right-handedness. Their report was affirmed by a number of gays who reported a "high rate of left-handedness" in their community. Geschwind and Galaburda admitted that there had been no systematic examination of these claims. However, there was theoretical support for this connection based on animal experiments that showed "permanently low free testosterone levels and homosexual behavior" in the offspring of rats who experienced uterine stress.[6] Acknowledging the limitations of rat studies, Geschwind and Galaburda urged a systematic investigation of human populations to determine whether homosexuals were more likely to be left-handed than heterosexuals.

The connection between left-handedness and homosexuality was widely held, recalled professor Robert Williams in 2016. Left-handers were often characterized as deviants and homosexuals.[7] Despite popular perceptions, in the late 1980s and 1990s the few published human population investigations concluded that there was no relationship between left-handedness and homosexuality. Setting out to test Geschwind and Galaburda's speculations, a team from the neuropsychological clinic at the University of Texas at Austin compared self-reports of handedness and sexual preference in 89 university students selected from a pool of 2,500 pretested students, concluding that there was "no relationship" between handedness and homosexuality.[8]

Over the next decade, studies continued to find no relationship between left-handedness and homosexuality. For instance North Dakota psychologists B. A. Gladue and J. M. Bailey's investigation of a relatively large population of men and women reported that there were significant sex differences for "mental rotations and spatial perception, but not for handedness." And in fact, "none of these measures was significantly related to sexual orientation within either sex."[9]

The next year a retrospective review of 6,544 men gleaned from the Kinsey Institute for Research in Sex, Gender, and Reproduction for the years 1938 to 1963 found no difference in handedness between heterosexual and homosexual men. Nor did they find an "increased level of non-right-handedness in homosexual men."[10]

Indeed, by the century's end, all studies concluded that there was no statistically significant relationship between sexual preference and left-hand preference. As a result, researchers in the early twenty-first century would focus on the extent of non-right-handedness among gays as a proxy for handedness preference. In order to understand the significance of the new emphasis, two issues must be addressed. First, given the lack of a persuasive connection between handedness and homosexuality, why would researchers, with no new population data, continue to pursue what seems like a dead end? That is, what is the context in which this renewed interest takes place? Second, and connected to the first, is, what explains the rise in the focus on "non-right-handedness" as a category of analysis?

We begin with context. As we have seen throughout this book, studies of left-handedness have frequently been about something other than left-handedness itself. Generally, that "something else" involves much higher stakes than the role of handedness. Thus, historically, the left hand has served as a proxy to justify or authorize beliefs and interventions, such as segregating the sacred from the profane; identifying and categorizing mental illness; justifying racist and sexist beliefs. In the debate about the gay hand, the existence and identification of biological traits and mechanisms of homosexuality are at stake.

This can be seen in the dispute, as it reemerged in the 1990s, over the widely discussed and controversial claims of Simon LeVay that homosexuality resulted from neurobiological difference, especially the reduced density of cells in the hypothalamus.[11] The same stressors that putatively influenced hand preference, according to Stanley Coren, also caused homosexuality. "The finding of more left-handed male homosexuals is what we would have expected if homosexuality was caused by some form of neural damage or other pathology." For Coren, left-handedness resulted from insults and stress, most likely during pregnancy or birth. "Homosexuality," asserts Coren, was most likely "another soft sign or a 'rare trait' that serves as a sign of stress or damage along with left-handedness." This explained for Coren why so many investigations have discovered "an overabundance of left-handers among male homosexuals." He also argues that female homosexuality is tied to handedness, but in a different manner, with "a very high percentage of non-right-handers in the group of female homosexuals." He pointed to another Canadian study that found that 69 percent of a group of lesbian women were "either left-handed or weak right-handers—four times greater than among female heterosexuals." For Coren, that finding confirmed the "hormonal imbalance theory and the neuropathology explanations of homosexuality [that] predict increased left-handedness among lesbians."[12]

By the beginning of the twenty-first century the theoretical case for the connection between homosexuality and left-handedness was contradicted by the statistical evidence. Psychologist Clare Porac writes that persuasive research connecting homosexuality and left-handedness "is hampered by the low percentages of homosexuals and non-right-handers available for study."[13] In lieu of a single large population study, a meta-analysis provides an alternative research tool, because it combines the findings of a number of retrospective studies, producing a much larger population for examination than any single study. In a way meta-analyses are analogous to hedge funds, whose value can only be assessed by a careful analysis of their contents. Like the financial instruments contained within hedge funds, a study must meet specific requirements to be included in a meta-analysis. Unfortunately, similar to many hedge funds, the studies included in meta-analyses fall short of their presumed value. For a meta-analysis to be done properly, there have to be enough valid studies to bring together in the first place, and the data in each study must have been collected in a similar manner and must be defined the same way. Thus the terms "homosexual" and "heterosexual" must mean the same thing in all studies; definitions must be comparable or be able to be reconstructed in a way that translates into the same understanding of what is being measured. The designation of "left," "right," and in this case, "non-right-handed" must have been determined by similar and interchangeable methods. Of course, researchers are allowed to reinterpret the data from any of these studies as long as doing so can be reasonably justified. Only then can statistical analysis be performed.

Now, why did non-right-handedness replace left-handedness as a focus of research attention?

REEXAMINING THE TORONTO STUDY

Martin L. Lalumière, Ray Blanchard, and Kenneth Zucker of the University of Toronto's Centre for Addiction and Mental Health,[14] authors of the 2000 meta-analysis cited at the beginning of this chapter, were immersed in the debate about the biological etiology of homosexuality. Although sympathetic to LeVay's general claim that homosexuality resulted from neurobiological difference, their research also focused on birth order; they hypothesized that middle-born males were more likely to be homosexual than oldest and youngest males.

The Toronto researchers hypothesized that a meta-analysis would expose significant handedness differences that could provide "early neurodevelopmental determinants or correlates of sexual orientation, that is, a person's erotic preference for opposite-sex individuals (heterosexuality), same-sex individuals (homosexuality), or both (bisexuality)." They believed that studying handedness would provide "important and perhaps more direct information on the early neurodevelopmental basis of sexual orientation."[15] Thus, the putative left-handedness of gays provided an observable manifestation of what was assumed to be a useful proxy for atypical hemispheric laterality.[16]

The Toronto team reexamined twenty previously published studies, which in combination provided them with a population of 6,987 homosexuals (6,182 men and 805 women) and 16,423 heterosexuals (14,808 men and 1,615 women). Based on this combined population the team reported that homosexual men and women had a 39 percent greater chance of being non-right-handed than heterosexuals. While gay males were more likely to be non-right-handed, ambidexterity was more common among lesbians.[17] It is essential to recognize that without the category "non-right-handedness," the meta-analysis would have had to conclude that there was no statistically significant connection between left-handedness and homosexuality.

The problems associated with the category of non-right-handedness were examined in chapter 3. As noted, there is no single agreed upon definition of non-right-handedness. "Non-right-hander" is not synonymous with the categories of strong or weak left-hander or right-hander. Rather, it is a category called upon to justify a conclusion when there is no statistically significant finding of left-handedness based on an inventory instrument such as the Edinburgh or Annett inventories. For Lalumière and colleagues, non-right-handedness included "nonconsistent right-handers," in contrast to "exclusive right-handers" and subjects "who tended to favor the left hand versus participants who tended to favor the right hand" (579).

Acknowledging the inherent problem of combining populations from studies that employed different recruitment criteria, the Toronto researchers indicated that for inclusion in their meta-analysis, studies had to have employed similar definitions of left-handedness, homosexuality, and heterosexuality. The classification "homosexual" was restricted to "self-declared homosexuals, middle-aged men who never married, [and] men diagnosed with HIV."[18] Handedness had to have been established by administration of a recognized instrument, such as, but not limited to, the Annett or Edinburgh surveys. Finally, to be included, a study had to identify the sex of all subjects (578).

The conclusion by Lalumière and his colleagues that homosexuals were more likely to be left-handed and non-right-handed confirmed the team's hypothesis "that sexual orientation in some men and women has an early neurodevelopmental basis." Although their study did not identify the mechanisms responsible for the sexual orientation of left-handers, the authors hypothesized three possibilities: "cerebral laterality and prenatal exposure to sex hormones, maternal immunological reactions to the fetus, and developmental instability."[19]

The take-home message of the Lalumière meta-analysis—that homosexual men and lesbian women were more likely to be left-handed and non-right-handed than people who were not homosexual or lesbian—was widely publicized despite the fact that the major finding was that left-handers and non-right-handers *combined* increased the association with homosexuality. If non-right-handers had been excluded, left-handers alone would have had no greater association with homosexuality than right-handers.

As Chris McManus had feared (see chapter 3), other researchers who failed to find gays and lesbians more likely to be left-handed nevertheless reported that "handedness was associated with sexual orientation."[20] For instance, University of London psychologists Mariana Kishida and Qazi Rahman reported that their 2015 meta-analysis of 900 gay and heterosexual men found that gay men who had "feminine scores" on the childhood gender nonconformity test were more likely to be "extremely right-handed."[21] Similarly, a 2014 study by psychologists from the University of British Columbia reported that "asexual men and women (those with no sexual attraction to persons of the opposite sex) were 2.4 and 2.5 times, respectively, more likely to be non-right-handed than their heterosexual counterparts."[22]

STATISTICAL PERSUASION

Lalumière and his colleagues subjected their meta-analysis population to a wide range of sophisticated tests to determine the rates of homosexual left-handedness. Their study is compelling from a statistical and methodological perspective. McManus, who was skeptical of a connection between left-handers and homosexuality, conceded that a larger population study might reveal different results.[23]

Like McManus, I was impressed with the sophistication of the paper's statistical methodology and comprehensive literature review. Troubling, however, were the team's aggregate data, based as they were on the supposition of a uniform definition of what constitutes homosexuality; the team's definition of left, right, and non-right-handedness; and their assumption that handedness is a valid proxy for hemispheric dominance and laterality.[24] Results failed to demonstrate a difference between left- and right-handers, forcing the researchers to rely on the vague category of non-right-handedness. Given these limitations it seems premature to identify, or even to suggest, as the Toronto team had done, the possible etiological factors that result in the gay hand.

Still, the meta-analysis appeared to be a careful investigation drawing on the best available data and on an examination of the interaction of environmental pressures and neurobiological mechanisms. However, a close examination of the studies that constituted the meta-analysis raises additional questions that cannot be ignored.

Many of the articles examined in the meta-analysis failed to support the Lalumière hypothesis; a number of them had concluded there was no relationship between left handers and homosexuality. Thus, the 1987 study by psychologists Rosenstein and Bigler concluded that there was "no relationship between handedness and sexual preference" in their subjects.[25] While the Lalumière meta-analysis cites the Rosenstein and Bigler article, Lalumière actually had relied on a 1993 article "re-analysis" that focused on non-right-handedness rather than left-handedness. The refocus on non-right-handers revealed that "nonexclusive heterosexuals were 3.13 times more likely to be nonright-handed than were exclusively heterosexual subjects (p < .03)."[26] Thus, without the problematic and nonspecific category of non-right-handedness, the data show no connection between homosexuality and left-handedness.

One of the largest handedness studies (2,083 subjects) used by the Lalumière team examines a wide number of variables, but sexual preference or homosexuality are not among them.[27] Lalumière et al. nevertheless include this study by arbitrarily classifying never-married males as homosexual and ever-married males as heterosexual, even though such a classification of sexual preference by ever-/never-married status is unreliable. Similarly, the Toronto team includes studies that rely on Kinsey data from the late 1930s to mid-1960s—notwithstanding that data's incompatibility with current definitions of homosexuality. Three of the sources included in the meta-analysis are unpublished studies: an undergraduate honors thesis, a British MA thesis, and a poster at a professional meeting.[28]

In contrast with the authors' claim that their criteria aimed at reliability, Lalumière and colleagues include widely different definitions of handedness, taken sometimes from self-reporting and at other times from just one sign of left-handedness, such as writing, which—as discussed in the context of forced hand-switching (in chapter 5)—is notoriously unreliable for determining handedness. As the team admitted, the studies they selected for analysis "used different scoring criteria" to determine left-handedness, right-handedness, and non-right-handedness. As a result, "the percentage of participants in a given category, such as left-handers," could not be averaged across studies" (579).

There are other hard-to-defend inclusions, such as Stellman and colleagues' 1997 study of left-handers admitted to a hospital. Ninety-one percent of the hospitalized left-handers were 45 years or older; criteria used to identify homosexuality were never-married (420) versus ever-married (5,615).[29] Another relatively large study (378 homosexuals and 278 heterosexuals) that identified left-handers using the Annett instrument reported that "no association was found between left-handedness and homosexuality, although there was an excess of left-handers in subjects who had been tested for HIV infection (irrespective of whether the test was negative or positive)."[30] What a close reading of these articles reveals is that there is no increased risk of left-handedness among homosexuals.

What explains the persistence of the connection between homosexuality and left-handedness beyond the need to find positive results from research on the road to tenure? Like so much of what we have learned in these pages, this association, despite being clothed in the methods and conclusions of current science, reveals a deep and persistent set of beliefs that since early human history has connected observable trait difference with pathology and has stigmatized left-handers as profane. Given the historic attitudes toward homosexuality, its connection with the stigmatized left-hand seems "natural."

This takes us back to where we started. What, if any, is the often observed but unproven correlation between learning disabilities, talent, and handedness/laterality? Although I do not provide a conclusive answer in chapter 11, I examine why this connection remains unresolved. An investigation of the putative role of weakened laterality in schizophrenia, autism, and creativity provides a starting point.

11

DISABILITY, ABILITY, AND THE LEFT HAND

Radical reorganization [of the brain] may well be not beneficial, and perhaps explains why anomalies of handedness and brain lateralisation are found in a wide range of conditions, including dyslexia, stuttering, autism, and schizophrenia. . . . Beneficial combinations of modules may occur more commonly in left-handers, [but] . . . there is no need to assume that left-handers, overall, will be better at a particular skill than right-handers.

CHRIS MCMANUS (2002)

There is almost no pattern of cerebral lateralization or handedness that has not at some time been mooted as the cause or correlate of one kind of developmental disorder.

DOROTHY V. BISHOP (1990)

After more than a century of investigations, the dispute over whether left-handedness and ambidexterity are signs of disability or seats of talent remains unresolved.[1] Cesare Lombroso's claim in the nineteenth century that mental deficits resulted from weak cerebral asymmetry was supported by late nineteenth-century theorists, including Gustave LeBon, Enrico Morselli, and even Robert Hertz's mentor, Émile Durkheim.[2] In contradiction, reduced laterality was also thought to be linked to talent. Hertz, for instance, drew on the contemporary writings and practices of the British Ambidexterity Culture Society, which advocated training children to use both hands equally in order to activate the potential of both hemispheres. The symmetry resulting from this training, its advocates argued, would produce more intelligent and gifted children.[3]

In the twenty-first century, investigators continue to explore whether left-handedness and reduced hemispheric laterality play a role in the etiology of deficits and talent—and, if so, what the role might be. Although in the West left-handedness is no longer viewed as evidence of mental and racial inferiority, it remains a prime factor under consideration to explain developmental and mental disorders. No longer measuring physical features and identifying suspect traits, researchers now use dichotic testing, molecular modeling, meta-analyses, and brain imaging. Relying on these new tools, today's researchers have reached conclusions similar to those of their early twentieth-century predecessors—that disruption of typical right-hand/left-hemisphere dominance increases the risk for mental illness.[4] Talent and creativity are assumed to emerge in a subset of this population that benefits from, but is not disturbed by, increased interaction and integration across both hemispheres.[5]

ARTICULATING THE ARGUMENT

Marian Annett, whose right shift (RS) theory was explored in chapter 8, articulated this argument at the turn of the current century, when she endorsed British psychiatrist T. J. Crow's hypothesis that failure to attain normal lateral dominance resulted in either schizophrenia or autism.[6] Annett asserted that her theory explained the relationship between disrupted asymmetries and schizophrenia and autism.[7]

Asymmetry, according to Annette, resulted from "*right hemisphere disadvantage*, rather than [left] hemisphere advantage" (198). Three hypothesized genes determined laterality: 1) *RS*–, the ancestral *primate allele** that is neutral for laterality; 2) *RS*+, the dominant human allele that carries instructions to *impair* the growth of the right hemisphere; and 3) *RS*+a, a mutant form of the *RS*+ that provides instructions to impair the growth of either the right or left hemisphere. Inheriting a combination of the *RS*+a and the normal *RS*+ can inhibit gene pruning of the right hemisphere. Inheriting two mutant alleles (*RS*+a and *RS*+a) can sometimes impair both hemispheres, resulting in weakened laterality that manifests itself as schizophrenia or autism.[8] Thus, in people with schizophrenia, both hemispheres resemble the right hemispheres of people without schizophrenia (208). In contrast, autism can result when a person inherits an agnosic gene (*RS*+a) from each parent, that codes for left or right at random (205).[9] In persons

---

* An allele, as we discussed in chapter 1, is an alternative form of a gene that resides on the same chromosome location.

with autism, "a random pattern of double hemisphere deficits would give scope for a range of developmental strengths and weaknesses as observed within the spectrum of autistic disorders." That is, people with autism lack self-mastery and have an inability to empathize—to imagine others' perspectives. They develop repetitive behaviors, are most comfortable with sameness, and are often unable to communicate or comprehend intention (209).

Annett argued that "alternative patterns of cerebral deficit would produce a range of possible consequences for cognitive functions, including the presence of one relatively normal hemisphere which might serve a special talent" (202). For instance, some people with autism display talents such as "drawing, music, or calculation" (197). Those diagnosed with schizophrenia and schizotypy, a condition with a number of schizophrenic signs and behaviors, have also been connected to reduced laterality and, as a result, heightened creativity. Writing in 1998 Swiss neurologists D. Leonhard and P. Brugger argued that the "unfocused semantic processing" typical in schizophrenia was also "characteristic of creative thinking" and "may constitute a selective evolutionary advantage allowing the genes predisposing to schizophrenia to proliferate despite the obvious disadvantages of this devastating disease."[10] Reviewing research since the late 1990s Australian psychologist Annukka Lindell has identified increased evidence that "schizotypy, creativity, and [reduced] laterality appear intimately related, implying a common, presumably genetic, underlying mechanism." Lindell cites "a cognitive bias toward broad processing" that enables insights "between disparate concepts and apparently unrelated ideas, [which] appears central to both the traits of schizotypy (e.g., magical ideation, perceptual aberrations) and superior performance on measures of creativity (e.g., divergent thinking)." This scenario, concludes Lindell, "appears to reflect predominant right hemisphere processing."[11]

Michael Corballis found Annett's hypothesis ingenious but ultimately unsatisfying, because it rested on "speculations, piled one on top of another, and so has the fragility of a house of cards." Corballis was not convinced that there was a single gene responsible for psychosis. He nevertheless accepted Annett's proposition that both schizophrenia and autism were connected to disruptions in cerebral dominance. Annett's hypothesis, wrote Corballis, "might well provide a useful starting point in the search for the genetic mechanisms underlying handedness, cerebral dominance for language, schizophrenia and autism, not to mention ... intelligence."[12]

Chris McManus was skeptical of Annett's hypothesis, arguing that his two-allele model could reach similar conclusions about the etiology of schizophrenia without having to be tied to lateralization.[13] McManus also found no evidence that, as Annett implied, extreme left- and right-handers were intellectually impaired. Nor was he persuaded that weak laterality was a feature of autism or that autism and schizophrenia are comorbid. Nevertheless, like Corballis, McManus conceded "there is almost certainly some relation of both schizophrenia and autism with atypical lateralization."[14]

Despite their differences, Corballis and McManus agreed with Annett that left-handedness itself was not a valid proxy for cerebral

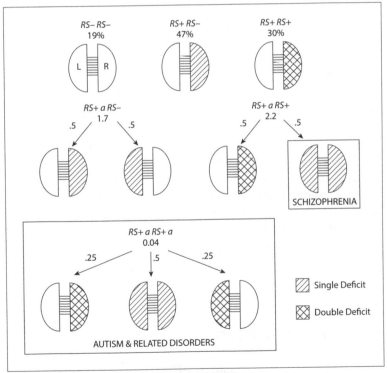

Representations of Annett's patterns of hemisphere deficits associated with each genotype. She hypothesizes that "alternative patterns of cerebral deficit" produce a spectrum of cognitive functions, from normal and talented to mental disorders. (Courtesy of Marian Annett.)

laterality. What role, she asked, does handedness play in determining mental illness? Handedness itself, she answered, was a *weak effect but not a cause* of schizophrenia and/or autism. Like Geschwind and Galaburda earlier, with whom she generally disagreed, Annett concluded that disrupted laterality, not left-handedness, was the condition that led to these disorders.[15]

SYMMETRY, ASYMMETRY, NON-RIGHT-HANDEDNESS, AND PSYCHOSIS

Two decades later there continues to be a steady stream of publications claiming that weak laterality and left-handedness increase the risk of schizophrenia and autism. Disregarding admonitions from Geschwind, Annett, Corballis, and McManus, most of these studies rely on handedness as a proxy for hemispheric laterality, in part because it is easier to determine handedness than cerebral dominance. The latter, at a minimum, requires brain scanning or intensive testing. Even so, as we have seen throughout this book, determining handedness can often be problematic. As important, none of the studies, including those discussed below, has been able to demonstrate that *left-handedness alone* is associated with either schizophrenia, autism, or creativity. In an attempt to overcome this limitation many studies combine rates of left-handers and "non-right-handers," which they variously define as mixed-handers who use either hand for the same task or who use different hands for different tasks. Such behavior, which can be identified in many humans, provides data and a rationale for publication, but strictly speaking it is irrelevant for identifying cerebral laterality. If these researchers reported that their substantive finding was that "left-handedness is unrelated to schizophrenia and autism," their chances of being published would be diminished.

SCHIZOPHRENIA AND LATERALITY

The meta-analyses of schizophrenia (like the one on homosexuality discussed in chapter 10) combine a number of published studies that determine laterality based on handedness. One of the most comprehensive of these is a 2001 meta-analysis of studies published in the

British Journal of Psychiatry by a Dutch team led by psychologist Iris Sommer. Examining 19 studies on handedness, 10 on dichotic listening, and 39 on "anatomical asymmetry" in people diagnosed with schizophrenia, the team reported that left-handers and non-right-handers *combined* had almost twice the risk of developing schizophrenia as nonschizophrenic controls.[16] If they had restricted their study only to left-handed schizophrenics, the authors would have been forced to admit that schizophrenics were *not* more likely to be left-handed.

Subsequently, the team focused on people with signs and behaviors associated with schizotypy. Finding no relationship with strong left-handers, or for that matter hand preference, they turned to non-right-handers who scored higher on schizotypy questionnaires than right-handers. Thus, assuming that non-right-handedness was a proxy for bilateral lateralization, they concluded that non-right-handers had a greater risk of schizophrenia than left-handers and right-handers.[17]

Other recent European and North American investigations have similarly relied on non-right-handedness as a proxy for hemispheric lateralization. Thus, a 2005 Australian meta-analysis found that "atypical hand dominance patterns were significantly greater in schizophrenia patients than in control subjects."[18] An Italian study confirmed "atypical handedness" and non-right-handedness among Italians with schizophrenia.[19]

Another Dutch team concluded that non-right-handed adolescents had an increased risk of psychoses and other mental illnesses.[20] This suggestion was reinforced that same year by an Australian team reporting "robust evidence that left- and mixed-handed children perform significantly worse in nearly all measures of development than right-handed children."[21]

In short, all of these studies failed to uncover a causative or other connection between schizophrenia and *left-handers*. But rather than emphasizing this finding, which was unlikely to be published, investigators fell back to making the connection to non-right-handedness, which had the advantage of vagueness and plasticity and most likely can be made to fit the behavior of a segment of any population.

Lately, some investigators, eschewing both meta-analysis and reliance on handedness, have turned to functional magnetic resonance imaging (fMRI) to determine hemispheric linguistic laterality. Using this technology, a French team recently sought to identify the relationship between hemispheric gray matter volume and right-hemisphere function. The investigators reported a reduction in gray matter volume in 27 right-handed schizophrenic patients in comparison with 54 nonschizophrenic right-handed controls, suggesting, as the other studies had, that schizophrenia was associated with weakened asymmetry.[22] However, this finding must be tempered by the fact that the schizophrenic patients in this study (in fact, all studies) were medicated with agents, including atypical antipsychotics, some of which have been shown to reduce gray matter volume.[23]

What this brief history suggests is that although there is a persuasive and logical hypothesis tying schizophrenia to weak laterality, the evidence for this connection remains weak, not least because it continues to be based on notoriously unreliable handedness proxies. There has been a slow but steady acknowledgment that only through investigating the brain directly, rather than through proxies, can the persistently assumed connection between schizophrenia and weak laterality be tested.

Which takes us back to Annett, who was convinced that both weak laterality and schizophrenia were caused by a mutation of a single gene, *RS+*. Implicit in her hypothesis, and in all of the studies examined in this section, is the assumption that schizophrenia is a single disease whose etiology can be "discovered." But as the German biopsychologist Sebastian Ocklenburg reminds us, "Schizophrenia is not a single disorder but a group of heritable disorders caused by different genotypic networks leading to distinct clinical symptoms." Thus, if one seeks to uncover a link between laterality and schizophrenia, the genetics should not be "mapped on schizophrenia as a whole but on discrete schizophrenia symptoms."[24]

Autism

Investigations into a possible link between autism and weakened asymmetry followed a trajectory similar to those described above for schizophrenia. For instance, in 2008 an Israeli team reported that children with coordination disorders comorbid with ADHD, similar to those diagnosed with autism, were more frequently left-handed or non-right-handed than a control population.[25] A 2015 meta-analysis by Norwegian researchers Anne L. Rysstad and Arve V. Pedersen reported, based on a combined sample population of 497, that the prevalence of non-right-handers and left-handers was greater within the autism spectrum disorder (ASD) population than in the general population. Their findings, wrote the authors, confirm the long-held association between ASD and left-handedness. Rysstad and Pedersen draw on the initial Geschwind testosterone hypothesis for evidence of the higher numbers of male left-handers. Although Rysstad and Pedersen were aware of the controversy over the validity of left-handedness as a proxy for hemispheric laterality, they insisted that their finding of a causal relationship between left-handedness and non-right-handedness and ASD was valid because theirs was the first to provide statistical evidence that non-right-handers were more common among those with ASD than among controls.[26] While this may be so, it also is irrelevant with regard to the validity of their finding.

Rysstad and Pedersen cited the recent meta-analysis by Emory University's Jessica Preslar and her colleagues (of which I am one), which tested the hypothesis that there was "an association between autism and laterality that would be moderated by handedness, sex, age, brain region studied, and level of autism." We began with 259 papers indexed in the PubMed website since 2000 that made this link but quickly narrowed the analysis to the 54 papers that elaborated on their methodology, including controls and reliable and accepted criteria for the definitions of autism, handedness, and asymmetry.[27] Additionally restricting the analysis to studies that measured hemispheric laterality directly, using scanning or other appropriate tools, reduced our sample to four papers. From these, wrote Preslar, we "found a moderate but non-significant effect size of the group on lateralization" (64).

While this finding was consistent with a "decrease in the strength of lateralization in the autistic group," further analysis of sex, along with a metaregression of handedness revealed that these variables were not significant in determining outcome (64, 82–88). The Preslar team also questioned the reliability of the other variables in these studies, including definitions of ASD, inclusion populations (some studies include only high-functioning persons with ASD, for example), age, sex of those studied, and the lack of imaging and other tools to determine hemispheric laterality (85–88). Finally, that all autisms have the same etiology and that all left-handedness and non-right-handedness produce the same effect remains to be demonstrated (64, 85–88).

Although the *theoretical* basis for assuming a connection between weak laterality and autism spectrum disorder is persuasive, the evidence remains weak. The same can be said of the claims that weak laterality plays a causative role in schizophrenia or creativity.²⁸ Despite the claims that schizophrenia and autism result from weak laterality, a plausible alternative explanation should be tested. That is, given the common finding that people on the autism spectrum have great difficulty with empathy, it seems that what they experience may be the result of hyperlaterality and hyperfocus, so that their ability to contextualize, which is more or less located in their nondominant hemisphere, is inhibited. Finally, it is possible that both handedness and reduced laterality are sometimes the effects rather than the causes of autism and schizophrenia.

The association between insanity and left-handedness has a long history, one that predates modern science by millennia. The longevity of this association does not in itself invalidate the more recent claims discussed above, but it does force us to reexamine the extent to which claims based on technology, methodology, and ideology are ultimately any more sustainable than earlier claims. We must always examine the degree to which seemingly objective explanations are reinforced by subterranean beliefs.

As important, most twenty-first-century studies continue to rely on the discredited assumption that handedness is an appropriate proxy for linguistic laterality. As has been repeated throughout this book, most left-handers are not linguistically right-brained, nor do they represent a majority of those who are weakly lateralized. Only with the addition of so-called non-right-handers or mixed-handers can left-handers be characterized as having a higher risk of mental illness than right-handers. Nevertheless, the majority of those with right-hemisphere linguistic dominance, even when adjusted for population at risk, are right-handed, as are the vast majority of those diagnosed with schizophrenia and autism.

What accounts for the persistence of the belief, despite evidence to the contrary, that left-handers and mixed-handers are more likely to develop mental illness than right-handers? This assumption has its origin in the long-held human belief that physical and motor "abnormality" must have behavioral consequences. When we examine our current theory, we assume that objective science, rather than cultural values, informs our conclusions. But history teaches, without exception, that values can never be separated from scientific findings. For this reason, science demands that claims always be tentative, and the scientific method insists that all theories be subjected to confirmation—that they be what the philosopher Karl Popper called "falsifiable." There is nothing wrong with theory so long as its tentative nature is acknowledged. Creating and testing hypotheses are the legitimate and necessary tasks of scientists. But, warns Chris McManus, "evolutionary theories are easily spawned and notoriously difficult to refute." McManus endorses the quip of the distinguished British biologist Steven Jones that "evolution is to allegory as statues are to birdshit. It is a convenient platform upon which to deposit badly digested ideas." For McManus, handedness genes "should be very similar to the genes that determine the side of the heart in vertebrates. For the moment," he concludes, "we must be content to wait for those genes to be identified."[29]

Researchers are under great pressure to produce positive, that is publishable, results. As psychologist D. V. Bishop noted a quarter of a century ago, "there is almost no pattern of cerebral lateralization or handedness that has not at some time been mooted as the cause or correlate of one kind of developmental disorder." Bishop had suggested three approaches to resolve these claims: awareness of the methodological problems that result in specious associations; examining competing hypotheses to determine reasons why and how handedness and disabilities might be linked; and finally, developing assessment procedures.[30] Applying these criteria over the following decades, Bishop found no persuasive connections between learning disabilities and handedness. This has led her to suggest that language ability influences cerebral lateralization rather than the other way around.[31]

Whether one agrees with Bishop's hypothesis or not, her wider findings of a lack of robust connection between handedness and learning disorders are persuasive. In terms of the studies examined in this chapter, achieving positive results depended on merging the often vague category of non-right-handedness with left-handedness. Without what might facetiously be called a sleight of hand, there would have been no unambiguous findings connecting left-handers with mental illness. This tactic of characterizing non-right-handers as a type of left-hander enabled research into the supposed pathology of left-handers to continue, when a focus exclusively on left-handers would have eliminated this line of research.

Conclusion
DOES LEFT-HANDEDNESS MATTER?

Left-handedness is an enigma worth unraveling. Handedness research in the twentieth century produced not only promising pathways of understanding but also a number of dead ends. The heavy dependence on incidence rate studies produced extended controversies and rampant inconsistencies in the attempts to connect left-handedness to both pathological and non-pathological traits. . . . A post on a Facebook page . . . asked the question What if left-handed people are actually normal? The answer to that question is that they actually are normal.

CLARE PORAC (2016)

Albert Galaburda, now the Emily Fisher Landau Professor of Neurology and Neuroscience at Harvard Medical School, concludes that even after decades of investigations using the latest methodologies and technologies, the etiology of left-handedness remains a mystery. We still cannot explain why some people are left-handed and others right-handed. Nor can we explain why more males than females are left-handed. Unable to replicate its claims, Galaburda no longer subscribes to the GBG hypothesis that he and Norman Geschwind put forward in the mid-1980s. He rejects the initial hypothesis that intrauterine testosterone effects result in a cascade of illnesses and left-handedness, but rather thinks testosterone modulates plasticity, resulting from any of several developmental effects both genetic and nongenetic.[1] This evolution in attitude toward the GBG hypothesis is evident among other former Geschwind and Galaburda colleagues and students, including Kenneth Heilman, James E. Rooks Distinguished Professor of Neurology and Health Psychology at the University of Florida. "The concept of handedness," writes Heilman, "is a very complex one and there are many motor asymmetries that differ between people and these asymmetries often have different brain mechanisms."[2]

Likewise, Galaburda now is persuaded that there are multifactorial and yet to be identified influences that determine whether and to what extent a person is left-, right-, or mixed-handed. Although he and Geschwind earlier were suspicious of genetic explanations of left-handedness, Galaburda embraces twenty-first-century genomics as offering the best tools and methods for unraveling the still elusive etiologies of handedness.[3] Similar to William Brandler and his colleagues (whose studies were examined in chapter 1),[4] Galaburda believes that a good place to start looking for answers is with the genes that code for cilia, hairlike projections on the cell wall from eukaryotes to humans. Because cilia are lateralized left, researchers hypothesize that these same genes in humans determine asymmetries such as organ placement and handedness.

So far, however, despite the sophisticated genetic tools employed, none of these investigations has been conclusive. Meanwhile, environmental explanations, including child rearing and early trauma, persist, cloaked in the vagueness of "epigenetics."[5] Investigators continue to explore possible selective advantages for the persistence of left-handedness, including the resurrection of the long-disputed and regularly dismissed claim that left-handedness remains advantageous for combat as long as the numbers of left-handers remain low enough to sustain surprise.[6]

This hypothesis, like all others, assumes that there has been and is a common definition of what constitutes a left-hander, a right-hander, and an ambilateral. But, as we have seen throughout this book, this is not so. Indeed, the primary obstacle to identifying the etiology of handedness is that there is not now nor has there ever been agreement on the definition of left-handedness, despite the variety of tools and hypotheses used to measure it. If we cannot agree on who and what constitutes a left-hander, it is impossible to identify its cause and its consequences.

Nor is there agreement about at what age or developmental stage handedness is determined. Thus, many investigators have assumed that a child's handedness is set at approximately her or his third year. However, genetic research and genetic modeling assume that the handedness gene (or genes) is inherited. Thus, in theory, the handedness gene (or genes) should be identifiable in utero. However, the populations that have been examined for determining handed prevalence are almost always at least school age or older, with the vast majority generally much older. Complicating this problem is that children historically have been and—on most of the planet—continue to be forced to use their right hands for eating, writing, and other major motor tasks. Although these pressures are not always effective, they nevertheless distort the rates of "natural" left-handers. As a result, whichever instrument is adopted to measure the left-handed population, it has a substantial margin of error built into it, so much so that it calls into question claims of statistical significance. This problem was elaborated in chapter 3, which examined the unbelievably low prevalence of left-handers reported in China and India. Even in the United States and much of the West, where left-handers are no longer forced to switch, the practice remains strong among recent immigrant populations.

The histories of left-handedness explored in this book suggest that there is no single cause of left-handedness. Instead, we should view left-handedness as a syndrome, that is, as a collection of signs (measurable) and symptoms (self-reported). Although Coren titled his book *The Left-Hander Syndrome,* he defines *syndrome* as synonymous with *disease*,[7] rather than, as elaborated in chapter 1, a distinct clinical entity based on a combination of signs and symptoms whose etiology has not been identified. Similar signs and symptoms can have different causes—for example, pneumonia may result from a variety of etiologies. And the same or similar causes can produce very different presentations. Given this perspective, we cannot simply assume that subjects with similar responses to surveys or with similar self-reports share the same handedness etiology. Nor should we assume that those who do not fit these criteria have different etiologies from those who do. The histories and persistence of etiological claims reveal that it is perfectly plausible for some persons to fit the syndrome for genetic reasons, while others became reliant on the left hand due to trauma and still others rely on the left hand because of educational or child-rearing pressures.

A number of researchers have suggested multiple etiologies that separately might account for left-handedness—including Margaret Clark in the 1950s, G. C. Ashton and Michael Corballis in the 1980s, Murray Schwartz in the 1990s, and most recently Dorothy Bishop.[8] While acknowledged, these suggestions have not been taken up by others. There are a number of obstacles to a sustained multi-causal exploration for left-handedness, not least being that such an acknowledgment would call into question handedness data, because researchers could not assume that a population of left-handers shared a common cause. This alone would problematize most claims about the etiology of left-handedness, not to mention theories about the relationship between handedness, learning disabilities, and talent. As we have seen, investigations of handedness have been driven by disciplinary training, tools, practices, and, lately, funding. Moreover, researchers have generally used handedness as a proxy for wider concerns, including demonstrations of the validity of their theories, disciplinary practices, and methods.

More than a quarter century ago Norman Geschwind and Albert Galaburda insisted that many statistical examinations were flawed because the data they examined were not reliable. Their hypothesis assumed that only some left-handedness and some anomalous dominance resulted from uterine trauma. Their critics insisted that Geschwind and Galaburda's claims were invalid because when examined, populations of left-handers were not statistically significant; that is, less than 95 percent met the criteria. But if there were a variety of causes of left-handedness, the statistical significance test would be irrelevant unless the populations could be separated by cause. But the most credible critic of the Geschwind and Galaburda hypothesis was Albert Galaburda himself. Their hypothesis, he insisted, was good for the time, but, he admitted, "we were wrong," as were the theories of opponents and critics. We know things today, according to Galaburda, that we couldn't have known then; we have tools that had not existed then.[9]

Finally, and most important, is the putative connection between left-handedness, learning disabilities, and talent. Here, the assumption that left-handers are linguistically right-brained while right-handers are left-brained is false.

I believe that clues to answering the questions raised in this book reside in the histories of left-handedness. The history of left-handedness parallels that of other disabilities. The "profane" left hand has historically been opposed to the "normal," often sacred, right hand. This antagonism was informed by both transcendent and culturally specific beliefs. We have seen that the damage produced by discrimination against left-handers was greater than the supposed pathology resulting from left-handedness. By the early twentieth century the stigmatizing practices toward left-handers was replaced by scientific, social, and educational theories that (re)authorized forcing left-handers to become right-handed. Although forced hand-switching is less frequent in the West, it continues to be widely practiced among the vast populations of China, India, and much of Africa. Moreover, the assumption of so many researchers that left-handedness is connected to learning disabilities has contributed to the persistent belief that left-handers are abnormal.

In addition, the continued production of studies combining left-handers and non-right-handers reinforced the assumption that left-handers were "abnormal," while it sanctioned practices aimed at normalizing their behaviors. However well-meaning, this desire to create conformity authorized the forced retraining of left-handers. Although theories implicating a greater risk of mental illness among left-handers remain highly speculative, the harms resulting from "curing" left-handers have been well documented.

These practices, along with other discrimination toward left-handers have lately been subjected to intense challenges by left-handers themselves, similar to those that earlier emerged in the wider disabilities movement. The reaction of left handers to perceived discrimination has taken a number of paths. Most recently it has manifested itself in websites worldwide, where left-handers post their experiences (generally negative) and reinforce a widely held belief among left-hander activists of the superiority of lefties in intellectual tasks. The evidence for the latter invariably includes lists of famous and talented left-handers, such as *Time* magazine's 2014 list of "Top 10 Lefties": Barack Obama, Bill Gates, Oprah Winfrey, Babe Ruth, Napoleon Bonaparte, Leonardo da Vinci,[10] Marie Curie, Aristotle, Ned Flanders (a fictional character from the Simpsons), and Jimi Hendrix.

There has been an exponential increase in websites and retail businesses catering to the desires of left-handers for implements and environments suited to their needs. International Left-Handers Day (August 13) was established in 1976 "to bring attention to the struggles which lefties face daily in a right-handed society." Left-Handers Day has grown from a curiosity to an annual international celebration reported extensively in the press and media.[11]

As a "left worker" I receive numerous communications from left-handers, most recently from Australia, North America, Britain, France, and Germany. What unites these communications is the recollection of humiliations suffered by left-handed children in schools in the 1940s to 1960s, a desire for acknowledgment of persecution, and recognition that left-handers have special talents.[12] Chapter 4 reproduced testimonies by left-handers from Europe and North America elaborating their maltreatment as children. Their intense residual anger, now often morphed into pride, is summed up in a 2016 posting by a 51-year-old French woman, Rosette. Her torture, Rosette remembered, began in kindergarten, where "my teachers forced me to use my right hand by making me put my left hand behind me and, if I used my left hand, they hit me on the knuckles with a ruler!" Angered by the memory of her treatment, Rosette is determined that her son will not suffer a similar fate. To that end she has forcefully insisted that her son's teachers "not interfere with his left-hand preference." Even now in 2016, she writes that although "it seems unlikely yet! we are in a world organized for righties, because they are more numerous," but "not more talented!"[13]

We need not exaggerate the adverse consequences of retraining to recognize how these humiliations reflected wider cultural practices that also rely on the notion of normality to authorize sexism, racism, segregation, homophobia, and xenophobia. To the contrary, Robert Hertz insisted that the liberation of left-handers would encourage wider cultural diversity. The evidence for this is not limited to the developed world. As the comparison in chapter 3 of the culturally conformist Temne of Sierra Leone with the more permissive Arunta aboriginals of Central Australia demonstrated, societies, like the Temne, that discriminate against left-handers are also less tolerant in general.[14]

Wherever left-handedness has been liberated from ancient prejudices, its reported numbers appear to have increased. The exception that may prove the rule is the higher prevalence of left-handedness among African Americans compared to European Americans discussed in chapter 3. Slavery, segregation, and its sequels denied the humanity of African Americans as it robbed them of human freedom. One ironic result seems to be that no one in the dominant society paid much attention to whether a slave or their descendants wrote (which during slavery was generally forbidden) or ate with their left hand. Although the data are admittedly limited, a reasonable hypothesis is that there was much less pressure on African Americans to conform to the general bias against left-handers. What we do not know is the influence of slave and African American culture on African American attitudes and practices toward left-handers. If we can obtain better data, left-handedness among African Americans may provide a more accurate measure of the natural distribution of handedness than we often possess.*

Although my evidence admittedly is anecdotal, I have encountered many more left-handers than the standard 10 to 12 percent. Perhaps this is because, as *Washington Post* reporter Michael Cavna wrote, left-handers "have a definite radar for each other."[15] But I also am persuaded that, as my German correspondent Michael Hackh insists, the number is larger but often hidden, because many left-handers are unaware that they have been converted. And, again anecdotally, in the groups where I hang out, university faculty and students, the percentage of left-handers often exceeds 50 percent. As to whether they are smarter or more talented—concepts that are themselves culturally freighted—I remain agnostic.

Although the histories of left-handedness that make up the chapters of this book do not in themselves reveal the etiology of handedness, they do provide a platform for rethinking approaches to—for a fresh examination of—this mystery. Despite the great intelligence and skill of current researchers, I suspect that until they acknowledge the likely multifactorial etiology of handedness, we will come no closer to solving the mysteries of handedness than those who have preceded us.

---

*Testing this hypothesis requires more data than are currently available. One reason for this has been the exclusion of African Americans from research populations. For instance, many studies of left-handedness drew on elementary and secondary school pupils, but because of segregation and other prejudices African American children were rarely included in the populations examined.

Acknowledgments

I am indebted to many people and institutions that have supported this project. Although left-handedness has been a lifelong interest, the impetus for my academic investigation about the left hand was reawakened by my observation of pediatric Tourette syndrome and attention deficit patients at Brown University's Memorial Hospital of Rhode Island in the mid-1990s. There, thanks to the encouragement of the clinic's directors, Louise Kiessling and the late Joseph Hallett, I began my first systematic examination of handedness, which raised more questions than it answered. My work at that time was supported by a National Institute of Humanities Collaborative Grant that freed me from my duties at San Diego State University (SDSU), where I was professor of the history of medicine. My SDSU colleagues, some of whom have endured decades of my speculations about the left hand, include Stephen S. J. Roeder, the late Michael Carella, Ross Dunn, Joanne Ferarro, Steven Colston, and William Weeks.

When I returned to San Diego from Providence in 1996, I affiliated with colleagues in neuropathology and pediatric infectious disease at the University of California at San Diego School of Medicine (UCSD), where among other things I attended the weekly brain-cutting seminar in the morgue of the VA hospital. There, while learning brain anatomy and pathology, I became a friend, colleague, and eventually, a collaborator with neuropathologist Henry (Harry) Powell, who would encourage me in my left-handed research for the next two decades, and who read the semifinal book manuscript, offering valuable insights. Although I left San Diego for Emory University in 2000, I continued my collaborations with UCSD faculty.

My appointment at Emory as the Nat C. Robertson Distinguished Professor of Science and Society provided generous research support. At Emory I was invited to join the Department of Behavioral Science and Health Education (BSHE) by its chair, Claire Sterk, who also included me in some of her many projects on addiction, self-medication, and behavior. I appreciate the support I received from Dean Robin Forman, Senior Associate Dean Michael Elliott of Emory's College of Arts and Sciences, as well as from Jim Curran, Dean of Emory's Rollins School of Public Health, and Senior Associate Dean Richard Levinson.

Eventually, I would also hold appointments in Neuroscience and Behavioral Biology and the Program of Human Health. Because of my joint appointments I was in a unique position to pursue my interests across traditional institutional divisions. Although I often was uncertain where I belonged, my colleagues in all these departments and programs welcomed me. Special thanks to my department chairs Michael Windle, Paul Lennard, Michelle Lampl, and Kevin Corrigan, all of whom encouraged this project by demanding less than I owed in service.

I thank Colin Talley, my colleague in Public Health and dependable friend, critic, and collaborator for more than three decades. Arri Eisen, along with Paula Frew, Randy Packard, Peter Brown, and James Dobbins (at CDC), recruited me to come to Emory. I hope they have no regrets, although Randy Packard left for Johns Hopkins soon after I arrived.

I appreciate the support of other Emory University colleagues, including Linda Cendales, Deanne Dunbar, Mary Horton, Gary Laderman, Delia Lang, Philippa Lang, Robert McCauley, Laurie Marino, Andy Nahmias, Clyde Partin, Cynthia Patterson, Bradley Pearce, Elaine Walker, and Kenneth Walker. The late Dana White provided me with an array of sources about left-handedness in US sports and popular culture. Kevin Corrigan provided me with Plato's positive views of left-handers. I thank Laura Otis, who read early versions of my book manuscript and made numerous valuable suggestions. I appreciate the encouragement from and conversations with Sander Gilman, whose breadth and depth of knowledge is unmatched. I am deeply indebted to Alison Adams, associate director of Emory's Center for Faculty Development and Excellence, for being a dedicated supporter and disseminator of my left-hander project, including her production of an iBook *Talent and Deficit: The Anomalies of Left-Handedness* (Academic Exchange, Emory University, 2012).*

At Emory I was blessed with outstanding PhD students, a number of whom are now professors themselves. Harold Braswell was a valuable collaborator on issues of mental illness and gender. Claire Clark read my entire manuscript and made numerous valuable suggestions. Melissa Creary's dissertation about the porous cultural borders of race and color influenced my thinking about the meanings and definitions

---

* Available at http://itunes.apple.com/us/book/academic-exchange supplementary /id556002148?mt=11.

of left-handedness. Shlomit Finkelstein brought her wide interest and knowledge of neuroscience to my graduate seminars. Les Leighton joined me in demonstrating that the history of medicine is a distinct discipline. Jennifer Sarrett reminded me that the stigma of mental illness could only be understood in cultural context. Jessica Preslar's honors thesis on autism and left-handedness was published in the journal *Laterality,* of which I am proud to have been the second author. Special thanks to my other graduate and undergraduate Emory students, who worked on various aspects of this project, including Daniel Coppeto, Arsalan Derakhshan, Jillian Kinton, Daniel Kirslis, Sasha Klupchak, Grace Kwon, Aukje Lamonica, Hayley McCausland, Tiffany Petrisko, Jessica Root, Eve Shapiro, Apoorwa Thati, Natalie Turrin, Christina Tzeng, Erin Vinovskis, Eddie Wang, Rachel Weitzenkorn, Sarah Whited, Amanda Wight, and Clara Wynn.

John Krige of Georgia Tech, renowned scholar of the history of science, provided his insights, criticisms, and especially his friendship during our long Atlanta adventure. A very special thanks goes to Lydie Mepham of Atlanta and Paris for her amazing expertise and suggestion about left-handedness in China. Also, I am especially indebted to the renowned biologist Bruce Levin, who has been a friend, colleague, and interlocutor since my first days at Emory, and to Adriana Juncos, who always asked good questions.

Over the years I have had numerous professional exchanges and shared ideas about handedness and laterality with a group of outstanding and generous scholars worldwide whose work and suggestions have framed my understanding of the complexities of left-handedness and laterality. Some of these scholars have become the subjects of my study. These include: Marian Annett, Sheldon Benjamin, Lisa Bob, Maria Teresa Broncaccio, Michael Corballis, Tabea Cornel, Vincent Felitti, Stanley Finger, Roger Freeman, Albert M. Galaburda, Jeremy Greene, Katja Guenther, Michael Hackh, Sidney Halpern, Anne Harrington, Lauren J. Harris, Kenneth Heinman, Annukka Lindell, Marjorie Lorch, Chris McManus, Erving Polster, Clare Porac, Roy Resnikoff, Walter Schalick, Barbara Schildkrout, Mindy Schwartz, Donald Stein, Cynthia Truant, John Waller, Marlie Wasserman, Alice Wexler, Robert Williams (pseudonym), and Fabio Zampieri.

When I returned to San Diego from Atlanta in 2015, I was invited to join the Laboratory for Comparative Human Cognition in the Department of Communication at UCSD, where I am a visiting scholar.

I especially thank Stefan Tanaka for inviting me, and Angelia Booker and Michael Cole for keeping me in the conversation. I also am grateful to the SDSU history department, especially its chair, Joanne Ferraro, for welcoming me back and for providing me with an office and other valuable support.

My neighbor Carl Larsen, a retired newspaper editor, travel writer, and friend, walks with me every morning. Carl has made valuable suggestions as he endured my latest hypotheses and interpretations of new data. My cousin, Evan Batoff, supplied me with the handed preferences of his branch of my mother's family.

I am profoundly grateful to Jackie (Jacqueline) Wehmueller, my editor at Johns Hopkins University Press, who truly has been my collaborator. Beginning with a conversation at the May 2013 American Association for the History of Medicine in Atlanta, Jackie challenged, shaped, and reorganized my ideas. Numerous scholars of the history of medicine have proclaimed Jackie's editorial skills, intellectual depth, and grace, all of which I was very fortunate to experience. Although she is responsible for the title, I assume all responsibility for any weaknesses in the final text.

Also at Johns Hopkins University Press, I appreciate the thorough, persistent, and sensitive copyediting of Michael Baker. Catherine Goldstead was extremely helpful in, among other things, my obtaining permissions for the book's illustrations. I also received valuable input from managing editor Juliana McCarthy and senior acquisitions editor Matt McAdam. Mary Lou Kenney guided the production with grace.

Some of the material in this book, though in different form and context, appeared in earlier publications and is used with permission: "Cesare Lombroso and the Pathology of Left-Handedness," *Lancet* 377 (2011): 118–19; "Retraining the King's Left Hand," *Lancet* 377 (2011): 1998–99; "Retraining Left-Handers and the Aetiology of Stuttering: The Rise and Fall of an Intriguing Theory," *Laterality* 17, no. 6 (2012): 673–93; "Deficit or Creativity: Cesare Lombroso, Robert Hertz, and the Meanings of Left-Handedness," *Laterality* 18, no. 4 (2013): 416–36; and "Why Are There (Almost) No Left-Handers in China," *Endeavour* 37 (June 2013): 71–81.

I dedicate my book to the two left-handers who shaped my life in nonsinister ways: my mother, Gertrude Slotnikoff Klein (1919–2002), and my spouse, partner, and editor for more than half a century, Carol Rose Rubin Kushner.

Notes

Preface
Epigraph: M. C. Corballis, "The Genetics and Evolution of Handedness," *Psychol Rev* 104 (1997): 714–27.

1. H. I. Kushner, "Retraining Left-Handers and the Aetiology of Stuttering: The Rise and Fall of an Intriguing Theory," *Laterality* 17, no. 6 (2012): 673–93.

2. H. I. Kushner, *A Cursing Brain?: The Histories of Tourette Syndrome* (Cambridge, Mass.: Harvard University Press, 1999).

3. V. Llaurens, M. Raymond, and C. Faurie, "Why Are Some People Left-Handed?: An Evolutionary Perspective," *Philos Trans R Soc Lond B Biol Sci* 364 (2009): 881–94; M. Raymond and D. Pontier, "Is There Geographical Variation in Human Handedness?," *Laterality* 9 (2004): 35–51.

4. M. Barsley, *The Other Hand: An Investigation into the Sinister History of Left-Handedness* (Portland, Ore.: Hawthorne Books, 1967).

5. M. Barsley, *The Left-Handed Book: An Investigation into the Sinister History of Left-Handedness* (London: Souvenir Press, 1966); M. Barsley, *Left-Handed People* (unknown: Borden Pub Co, 1967); M. Barsley, *Left-Handed Man in a Right-Handed World* (London: Pitman, 1970).

6. S. Coren, *The Left-Hander Syndrome: The Causes and Consequences of Left-Handedness* (New York: Free Press, 1992), 92.

7. Ibid., 140.

8. I. C. McManus, *Right Hand, Left Hand: The Origins of Asymmetry in Brains, Bodies, Atoms, and Cultures* (London: Weidenfeld & Nicolson, 2002).

9. Two recent books, both in the Barsley tradition of demythologizing negative characterizations of left-handers, are Melissa Roth's *The Left Stuff: How the Left-Handed Have Survived and Thrived* (Lanham, Md.: Rowman & Littlefield, 2005) and Rik Smits, *The Puzzle of Left-Handedness*, translated from a revised and expanded 2010 Dutch edition of the original 1993 edition (London: Reaktion Books, 2011), an updated and revised translation of his 1993 Dutch study, *De Linkshandige Picador*. Lacking citations, Smits's book has limited value for scholars.

10. P-M. Bertrand, *Histoire de gauchers: Des gens à l'enfers* [History of Left-Handers: Upside down people] (Paris: Éditions Imago, 2001), 7–9. Here and in what follows all translation from French is mine.

11. Clare Porac, *Laterality: Exploring the Enigma of Left-Handedness* (London: Academic Press, 2016), ix, x.

12. A. K. Lindell, "Laterality: Exploring the Enigma of Left-Handedness" (in English), *Laterality* Epub (Sept. 8, 2016): 1–3, doi: 10.1080/1357650X.2016.1229326.

13. C. Kudlick, "Disability History: Why We Need Another 'Other,'" *Am Hist Rev* 108 (2003): 763–93.

Chapter 1. Genes and Kangaroos
Epigraphs: W. Tecumseh Fitch and Stephanie N. Braccini, "Primate Laterality and the Biology and Evolution of Human Handedness: A Review and Synthesis," *Ann NY Acad Sci* 1288 (2013): 70–85, 70; Andrey Giljov et al., "Parallel Emergence of

True Handedness in the Evolution of Marsupials and Placentals," *Current Biology* 25 (2015): 1878–84, 1878.

1. Cesare Lombroso, "Left-Handedness and Left-Sidedness," *North American Review* 177 (1903): 440–44; H. I. Kushner, "Cesare Lombroso and the Pathology of Left-Handedness," *Lancet* 377 (2011): 118–19.

2. G. LeBon, *L'homme et les sociétés, Leurs origines et leur histoire* (1881) (Paris: Jean-Michel Place, 1988).

3. Anne Harrington, *Medicine, Mind, and the Double Brain: A Study in Nineteenth-Century Thought* (Princeton, N.J.: Princeton University Press, 1987), 68–69, 88–89, 95–99.

4. L. M. Lansky, H. Feinstein, and J. M. Peterson, "Demography of Handedness in Two Samples of Randomly Selected Adults (N = 2083)," *Neuropsychologia* 26, no. 3 (1988): 465–77, 470.

5. M. L. Lalumière, R. Blanchard, and K. J. Zucker, "Sexual Orientation and Handedness in Men and Women: A Meta-Analysis," *Psychol Bull* 126 (2000): 575–92.

6. Robert Hertz, "La Prééminence de la Main Doite: Étude sur la Polarité Religieuse," *Revue Philosophique* LXVIII (1909): 553–80; H. I. Kushner, "Deficit or Creativity: Cesare Lombroso, Robert Hertz, and the Meanings of Left-Handedness," *Laterality* 18, no. 4 (2013): 416–36, 428–30.

7. A. K. Lindell, "Lateral Thinkers Are Not So Laterally Minded: Hemispheric Asymmetry, Interaction, and Creativity," *Laterality* 16 (2011): 479–98; M. Annett, "Schizophrenia and Autism Considered as the Products of an Agnosic Right Shift Gene," *Cogn Neuropsychiatry* 2 (1997): 195–214.

8. Rik Smits, *The Puzzle of Left-handedness* (London: Reaktion Books, 2011), 250.

9. M. C. Corballis, "The Genetics and Evolution of Handedness," *Psychol Rev* 104 (1997): 714–27.

10. Kushner, "Deficit or Creativity."

11. J. Preslar et al. "Autism, Lateralisation, and Handedness: A Review of the Literature and Meta-Analysis," *Laterality* 19 (2014): 64–95.

12. Honor Whiteman, "Left and Right-Handed Genes Discovered," *Medical News Today* (September 16, 2013), www.medicalnewstoday.com/articles/266093.php.

13. W. M. Brandler et al., "Common Variants in Left/Right Asymmetry Genes and Pathways Are Associated with Relative Hand Skill," *PLoS Gen* 9 (2013): e1003751, 1.

14. Although in most humans the heart is lateralized left and the liver is lateralized right, a surprisingly large number of humans (1 in 10,000) are lateralized the other way around. This condition is called *situs inversus*. I. C. McManus, *Right Hand, Left Hand: The Origins of Asymmetry in Brains, Bodies, Atoms, and Cultures* (London: Weidenfeld & Nicolson, 2002), 1–7.

15. The putative connection between left-handedness and dyslexia is complicated and controversial. As the handedness expert Marian Annett points out, "'phonological' dyslexics are less likely to be right-handed, while 'surface' or 'dyseidetic' dyslexics are more likely to be right-handed than the general population." M. Annett, "Dyslexia and Handedness: Developmental Phonological and Surface Dyslexias Are Associated with Different Biases for Handedness," *Percept Mot Skills* 112 (2011): 417–25, 417.

16. Brandler et al., "Common Variants of Left/Right Asymmetry," 7.

17. Ibid.

18. Although Brandler and Paracchini assumed that as in *situs inversus*, where the heart is lateralized left and the liver is lateralized right, this reversal of organs was accompanied by a reversal of handedness. Brandler et al., "Common Variants in Left/Right Asymmetry Genes," 6. It is not. Individuals with *situs inversus* are right-handed. McManus, *Right Hand, Left Hand*, 111–20.

19. W. M. Brandler and S. Paracchini, "The Genetic Relationship between Handedness and Neurodevelopmental Disorders," *Trends Mol Med* 20 (February 2014): 83–90.

20. J. A. Armour, A. Davison, and I. C. McManus, "Genome-Wide Association Study of Handedness Excludes Simple Genetic Models," *Heredity* (Edinburgh) 112 (March 2014): 221–25.

21. UPI Beta Science News, "Genetic Link to Being Left-Handed or Right-Handed Discounted: Interview with John Amour," in *Science News UPI Beta* 2013, https://www.sciencedaily.com/releases/2013/10/131001123943.htm.

22. M. Somers et al., "Linkage Analysis in a Dutch Population Isolate Shows No Major Gene for Left-Handedness or Atypical Language Lateralization," *J Neurosci* 35 (2015): 8730–36.

23. D. V. Bishop, "Cerebral Asymmetry and Language Development: Cause, Correlate, or Consequence?," *Science* 340 (2013): 1230531, 2; D. V. Bishop, *Handedness and Developmental Disorder* (Philadelphia: Blackwell Scientific, 1990).

24. Giljov et al., "Parallel Emergence of True Handedness in the Evolution of Marsupials and Placentals," 1878–84.

25. Ibid. A subsequent study of wallaby forelimb preference claims has replicated "previous research reporting a left-bias in forelimb use for different types of activities in macropod." C. Spiezio, B. Regaiolli, and G. Vallortigara, "Motor and Postural Asymmetries in Marsupials: Forelimb Preferences in the Red-Necked Wallaby (Macropus rufogriseus)," *Behav Processes* 128 (2016): 119–25.

26. M. C. Corballis, "Laterality and Human Evolution," *Psychol Rev* 96 (1989): 492–505, 496–97.

27. Emily Willingham, "Kangaroos Join Ranks of Animals Reported to Have Autism," *Forbes*, June 21, 2015, www.forbes.com/sites/emilywillingham/2015/06/21/kangaroos-join-ranks-of-animals-reported-to-have-autism/. Although they lack a corpus callosum, the hemispheres of marsupials and monotremes are connected and communicate through the anterior commissures.

28. Preslar et al., "Autism, Lateralisation, and Handedness."

29. Jadon R. Webb, Mary I. Schroeder, Christopher Chee, et al., "Left-Handedness among a Community Sample of Psychiatric Outpatients Suffering from Mood and Psychotic Disorders," *SAGE Open* (October 2013): 3.

30. Alexandra Sifferlin, "The Connection between Left-Handedness and Schizophrenia," in Time.com, November 2013, http://healthland.time.com/2013/11/01/the-connection-between-left-handedness-and-schizophrenia/.

31. "Left Handers and Intelligence," in Anything Left-Handed, 2013, http://www.anythinglefthanded.co.uk/.

32. Christopher S. Ruebeck, Joseph E. Harrington Jr., and Robert Moffitt, "Handedness & Learning," *National Bureau of Economic Research*, 2006 (working paper 12387).

33. Laura Clark, "Why Left-Handed Men Earn Five Percent More Every HOUR

Than Right-handers," *Daily Mail*, November 28, 2008, 2. Smits attributes this study to unnamed Johns Hopkins University researchers, writing that the study found a 15 percent difference, favoring left-handers. *The Puzzle of Left-Handedness*, 251. Unfortunately, because Smits's book lacks citations, it is impossible to validate his claims.

34. Zazzle, "Left-Handed T-Shirts," 2013, available from www.zazzle.com/left +handed+tshirts.

35. Betty Edwards, *Drawing on the Right Side of the Brain: A Course in Enhancing Creativity and Artistic Confidence*, 4th ed. (New York: J. P. Tarcher, 2012).

36. Melissa Roth, *The Left Stuff: How the Left-Handed Have Survived and Thrived* (Lanham, Md.: Rowman & Littlefield, 2005), 94.

37. Fitch and Braccini, "Primate Laterality and the Biology and Evolution of Human Handedness," 70–85.

38. S. Ocklenburg, C. Beste, L. Arning, et al., "The Ontogenesis of Language Lateralization and Its Relation to Handedness," *Neurosci Biobehav Rev* 43 (2014): 191–98; also see S. Ocklenburg, C. Beste, and O. Gunturkun, "Handedness: A Neurogenetic Shift of Perspective," *Neurosci Biobehav Rev* 37 (2013): 2788–93.

39. Bishop, "Cerebral Asymmetry and Language Development," 2.

40. McManus, *Right Hand, Left Hand*, 194.

41. Ibid., 197.

42. Ibid., 223–24. The corpus callosum is deeply implicated in these three disorders, but its exact role is poorly understood. S. Ocklenburg, Beste, Arning, et al., "The Ontogenesis of Language Lateralization and Its Relation to Handedness"; M. Bellani et al., "Laterality Effects in Schizophrenia and Bipolar Disorder," *Exp Brain Res* 201 (2009): 339–44.

43. R. E. Rosch, D. V. Bishop, and N. A. Badcock, "Lateralised Visual Attention Is Unrelated to Language Lateralisation, and Not Influenced by Task Difficulty— A Functional Transcranial Doppler Study," *Neuropsychologia* 50 (2012): 810–15.

44. Fitch and Braccini, "Primate Laterality and the Biology and Evolution of Human Handedness," 82.

45. H. I. Kushner, "Retraining the King's Left Hand," *Lancet* 377 (2011): 1998–99; H. I. Kushner, "Retraining Left-Handers and the Aetiology of Stuttering: The Rise and Fall of an Intriguing Theory," *Laterality* 17, no. 6 (2012): 673–93.

46. Milan Dragovic, "Towards an Improved Measure of the Edinburgh Handedness Inventory: A One-factor Congeneric Measurement Model Using Confirmatory Factor Analysis," *Laterality* 9 (2004): 411–19; M. Dragovic and G. Hammond, "A Classification of Handedness Using the Annett Hand Preference Questionnaire," *Br J Psychol* 98 (2007): 375–87; Stephen M. Williams, "Handedness Inventories: Edinburgh Versus Annett," *Neuropsychology* 5 (1991): 43–48; H. I. Kushner, "Why Are There (Almost) No Left-Handers in China?," *Endeavour* 37 (June 2013): 71–81.

47. Brandler et al., "Common Variants in Left/Right Asymmetry Genes and Pathways."

48. M. C. Corballis, *The Lopsided Ape: Evolution of the Generative Mind* (New York: Oxford University Press, 1991), 184. Also see M. C. Corballis, "Handedness and Cerebral Asymmetry: An Evolutionary Perspective," in *The Two Halves of the Brain: Information Processing in the Cerebral Hemispheres*, ed. K. Hugdahl and R. Westerhausen (Cambridge, Mass.: MIT Press, 2010), 65–88.

49. Bishop, "Cerebral Asymmetry and Language Development," 5–6.

50. McManus, *Right Hand, Left Hand,* 164–65; I. C. McManus, *Hypernotes to Right Hand, Left Hand: The Origins of Asymmetry in Brains, Bodies, Atoms, and Cultures,* 7:31, 13–14, www.righthandlefthand.com.

51. S. M. Scharoun and P. J. Bryden, "Hand Preference, Performance Abilities, and Hand Selection in Children," *Frontiers in Psychology* 5 (February 18, 2014): 82.

52. As a result, researchers such as Annett and McManus have constructed hypothetical genetic models of handedness assuming that they will reveal actual hand preference, but as creative as these models are they remain speculations. Corballis, *The Lopsided Ape,* esp.161–62, 227–29.

53. M. C. Sutter, "Assigning Causation in Disease: Beyond Koch's Postulates," *Perspectives in Biology and Medicine* 39 (1996): 581–92.

54. H. I. Kushner, *A Cursing Brain?: The Histories of Tourette Syndrome* (Cambridge, Mass.: Harvard University Press, 1999), 584–85; Ian Hacking, *Rewriting the Soul: Multiple Personality and the Sciences of Memory* (Princeton, N.J.: Princeton University Press, 1995), 8–38.

55. Roy Richard Grinker, "Autism and Culture: The Effect on Epidemiology and Diagnosis at Home and Abroad," in *Autism and Autism Spectrum Disorders: History, Diagnosis, Neurobiology, Treatment and Outcome* (London: Henry Stewart Talks, 2010); Terra Diane Ziporyn, *Nameless Diseases* (New Brunswick, N.J.: Rutgers University Press, 1992), 1–2.

56. Grinker, "Autism and Culture."

57. H. I. Kushner and L. S. Kiessling, "The Controversy Over the Classification of Gilles de la Tourette's Syndrome, 1800–1995," *Perspect Biol Med* 39 (1996): 409–35; H. I. Kushner, "The Cursing Patient: Neuropsychiatry Confronts Tourette Syndrome, 1825–2010," in *The Neurological Patient in History,* ed. Stephen Jacnya and Stephen Casper (Rochester: University of Rochester Press, 2012), 129–64.

58. Preslar et al., "Autism, Lateralisation, and Handedness."

59. Bishop, "Cerebral Asymmetry and Language Development," 5–6.

60. Margaret M. Clark, *Left-Handedness: Laterality Characteristics and Their Educational Implications* (London: University of London Press, 1957).

Chapter 2. Criminals or Victims?

Epigraphs: Abraham Verghese, *Cutting for Stone* (New York: Vintage Press, 2009), 148; Robert Parkin, *The Dark Side of Humanity: The Work of Robert Hertz and Its Legacy* (Australia: Harwood Academic Publishers, 1996), 65.

1. Mary Gibson and Nicole Hahn Rafter, "Introduction to Criminal Man by Cesare Lombroso," in *Criminal Man,* ed. Mary Gibson and Nicole Hahn Rafter (Durham: Duke University Press, 2006), 18; Mary Gibson, *Born to Crime: Cesare Lombroso and the Origins of Biological Criminology* (Westport, Conn.: Praeger, 2002), 104–7.

2. Robert Hertz, "The Pre-Eminence of the Right Hand: A Study of Religious Polarity" in *Death and the Right Hand,* ed. and intro. E. E. Evans-Pritchard, trans. Rodney Needham and Claudia Needham (New York: Routledge, 1960; reprinted 2004), 113; Robert Hertz, "La Prééminence de la Main Doite: Étude sur la Polarité Religieuse," *Revue Philosophique* LXVIII (1909): 553–80.

3. Gibson and Rafter, "Introduction to Criminal Man by Cesare Lombroso," 18; Daniel Pick, *Faces of Degeneration: A European Disorder, c.1848–c.1918,* Ideas in Context (Cambridge: Cambridge University Press, 1989), 104–7, 109–53.

4. Cesare Lombroso, *Criminal Man*, ed. Mary Gibson and Nicole Hahn Rafter, trans. Mary Gibson and Nicole Hahn Rafter (Durham: Duke University Press, 2006).

5. Cesare Lombroso and Guglielmo Ferrero, *Criminal Woman, the Prostitute, and the Normal Woman*, trans. Nicole Hahn Rafter and Mary Gibson (Durham: Duke University Press, 2004).

6. Cesare Lombroso, "Left-Handedness and Left-Sidedness," *North American Review* 177 (1903): 440–44, 442.

7. Ibid.

8. Robert A. Nye, "Heredity or Milieu: The Foundations of Modern European Criminological Theory," *Isis* 67 (1976): 335–55, 338; Robert A. Nye, *Crime, Madness, and Politics in Modern France: The Medical Concept of National Decline* (Princeton: Princeton University Press, 1984), 3, 4–5. Also see Gibson, *Born to Crime*; H. I. Kushner, "Deficit or Creativity: Cesare Lombroso, Robert Hertz, and the Meanings of Left-Handedness," *Laterality* 18, no. 4 (2013): 416–36.

9. Anne Harrington, *Medicine, Mind, and the Double Brain: A Study in Nineteenth-Century Thought* (Princeton, N.J.: Princeton University Press, 1987), 68–69, 145–46.

10. G. LeBon, *L'homme et les sociétés, Leurs origines et leur histoire* (1881) (Paris: Jean-Michel Place, 1988); Harrington, *Medicine, Mind, and the Double Brain*, 68–69. Chief among Lombroso's critics was the French sociologist and jurist Gabriel Tarde (1843–1904), who insisted that Lombroso's explanations failed to account for the social causes of criminality. Gabriel Tarde, *La Criminalité Comparée* (Paris: Librairie Félix Alcan, 1886; reprinted 1924); Nye, "Heredity or Milieu." To some extent the disputes over the validity of Lombroso's criminology reflected national rivalries and, according to historian Mary Gibson and criminologist Nicole Rafter, "recent scholarship has begun to reevaluate the place of Lombroso in a variety of contexts, including the histories of criminology, science, race, and sexuality." Gibson and Rafter, "Introduction to Criminal Man by Cesare Lombroso," 4–5.

11. LeBon, *L'homme et les sociétés*, 154. LeBon's influence on his contemporaries was enormous. É. Durkheim, *The Division of Labor in Society (1893)*, trans. W. D. Halls (London: Macmillan, 1984), 20–21; also see Howard I. Kushner, "Suicide, Gender, and the Fear of Modernity," in *Studies on Suicide and History*, ed. David Wright and John Weaver (Toronto: University of Toronto Press, 2009): 18–52, 32–33.

12. Harrington, *Medicine, Mind, and the Double Brain*, 88–89, 95–99.

13. Théodule Ribot, *Les Maladies De La Volonté*, 14th ed. (Paris: Félix Alcan, 1900); B. A. Morel, *Traité Des Dégénérescences Physiques, Intellectuelles Et Morales De L'espèce Humaine Et Des Causes Qui Produisent Ces Variétés Maladives* (Paris: J. B. Baillière, 1857).

14. Ian Dowbiggin, *Inheriting Madness: Professionalization and Psychiatric Knowledge in Nineteenth-Century France* (Berkeley: University of California Press, 1991); Mark S. Micale, *Approaching Hysteria: Disease and Its Interpretations* (Princeton, N.J.: Princeton University Press, 1995); Nye, *Crime, Madness, and Politics in Modern France*; Pick, *Faces of Degeneration*, 7.

15. Pick, *Faces of Degeneration*, 7.

16. Ibid., 122, 126. Lombroso's evolutionary views were greatly influenced by the Italian philosopher Giambattista Vico (1668–1744). According to Vico, the historical development of humanity was progressive. Early humans had no society

or spoken or written language. The only difference from "pure" animals and early humans, according to Vico, was the presence of a familial structure and burial rituals. Giambattista Vico, *The New Science of Giambattista Vico*, rev. translation of the 3rd ed. (Ithaca, N.Y.: Cornell University Press, 1968). Building on Vico, Lombroso assumed an evolution from the epoch of black races, to yellow races, and finally to the white race. His characteristics of each race corresponded with those of Vico: the black races were instinctive and superstitious, the yellow were organized in strictly aristocratic societies, while white races were characterized by democracy and reason. Lombroso's typological approach included the assumption that types remained immutable once they were formed. Lombroso's modified evolutionary theory is best described as what current evolutionary biologists have labeled "constitutional medicine," or holism. Fabio Zampieri, "Medicine, Evolution, and Natural Selection: An Historical Overview," *Quarterly Review of Biology* 84 (2009): 333–55; for Lombroso, see 337. For an overview of constitutional medicine, see S. W. Tracy, "George Draper and American Constitutional Medicine, 1916–1946: Reinventing the Sick Man" (in English), *Bulletin of the History of Medicine* 66, no. 1 (Spring 1992): 53–89. I thank Professor Fabio Zampieri, University of Padova, for having alerted me to Vico's influence on Lombroso's evolutionary thinking and for sharing his current work on Lombroso.

17. Lombroso, *Criminal Man*, 118–19.

18. Hertz, "The Pre-Eminence of the Right Hand," 94–95, 155, note 12.

19. Ibid., 95.

20. Elsdon Best, "Maori Nomenclature: Notes on the Consanguineous, Affinitative, Personal, Tribal, Topographical, Floral and Ornithological Nomenclature of the Maori Race of New Zealand," *Journal of the Anthropological Institute of Great Britain and Ireland* 32 (1902): 182–201.

21. Among Native Americans "the right hand stands for *me*, the left hand for *not me* and *others*." Raising the right hand symbolizes "bravery, power and virility," while the left hand is held lower than the right, symbolizing inferiority and "death, destruction, and burial." Ibid., 101.

22. Parkin, *The Dark Side of Humanity*, 2–3.

23. Ibid., 57.

24. Ibid., 16. Italics in the original. Unless otherwise stated, all emphasis in quotations is in the original.

25. Hertz, *Death and the Right Hand*, 86.

26. According to Hertz's wife, Alice, her husband's initial motivation for his essay on handedness was to advocate the advantages of ambidexterity for learning. Only subsequently, writes Hertz's biographer Robert Parkin, did Hertz view handedness as a social issue. Parkin, *The Dark Side of Humanity*, 3.

27. L. J. Harris, "Left-Handedness: Early Theories, Facts, and Fancies," in *Neuropsychology of Left-Handedness*, ed. Jeannine Herron, Perspectives in Neurolinguistics and Psycholinguistics (New York: Academic Press, 1980), 46.

28. John Jackson, *Ambidexterity: or, Two-Handedness and Two-Brainedness: An Argument for Natural Development and Rational Education* (London: Kegan Paul, Tench, Trubner & Co., 1905; reprinted 2009 by Bibliolife).

29. Ibid., xi.

30. Ibid., 19; Harris, "Left-Handedness," 47; also see L. J. Harris, "The Ambidextral

Cultural Society and the 'Duality of the Mind,'" *Behavioral and Brain Sciences* 8 (1985), 639–40. For a sample of the lecture presented at the society's meetings at this time, see "The Ambidextral Culture Society," *Lancet* 163, no. 20 (February 1904).

31. Hertz, "The Pre-Eminence of the Right Hand," 112–13.

32. Daniel Wilson, *The Right Hand: Left Handedness* (New York: McMillan, 1891).

33. Hertz, "The Pre-Eminence of the Right Hand," 113.

34. É. Durkheim, *Suicide: A Study in Sociology* (New York: Free Press, 1951), 241–50; H. I. Kushner and C. E. Sterk, "The Limits of Social Capital: Durkheim, Suicide, and Social Cohesion," *Am J Public Health* 95 (2005): 1139–43, 1140–41.

35. Gibson and Rafter, "Introduction to Criminal Man by Cesare Lombroso,"18; Gibson, *Born to Crime*, 104–7.

36. Lombroso, "Left-Handedness and Left-Sidedness," 444.

CHAPTER 3. BY THE NUMBERS

Epigraphs: M. C. Corballis, "The Genetics and Evolution of Handedness," *Psychol Rev* 104 (1997): 714–27, 714. Y. P. Zverev, "Cultural and Environmental Pressure against Left-Hand Preference in Urban and Semi-Urban Malawi," *Brain and Cognition* 60 (2006): 295–303, 296.

1. M. C. Corballis, "The Genetics and Evolution of Handedness," *Psychol Rev* 104 (1997): 714–27, 714.

2. C. Faurie and M. Raymond, "Handedness Frequency over More than Ten Thousand Years," *Proc Biol Sci* 271, suppl. 3 (2004): S43-45; Charlotte Faurie et al., "Variation in the Frequency of Left-Handedness in Traditional Societies," *Current Anthropology* 46 (2005): 142–47, 146; M. Raymond and D. Pontier, "Is There Geographical Variation in Human Handedness?," *Laterality* 9 (2004): 35–51.

3. S. Coren, "Measurement of Handedness Via Self-Report: The Relationship between Brief and Extended Inventories," *Perceptual and Motor Skills* 76, no. 3, pt. 1 (1993): 1035–42.

4. There is a fourth method—tests of skill for relatively less-practiced acts of the kind used by Marian Annett, Michael Peters, and other researchers. Because these are more time-consuming to administer, they have been used less frequently in estimating the prevalence of handedness subtypes. M. Annett, "Hand Preference and Skill in 115 Children of Two Left-Handed Parents," *Br J Psychol* 74 (1983): 17–32; M. Peters and P. Servos, "Performance of Subgroups of Left-Handers and Right-Handers," *Canadian Journal of Psychology* 43 (1989): 341–58.

5. R. C. Oldfield, "The Assessment and Analysis of Handedness: The Edinburgh Inventory," *Neuropsychologia* 9 (1971): 97–113.

6. M. Annett, "A Classification of Hand Preference by Association Analysis," *Br J Psychol* 61 (1970): 303–21.

7. D. Raczkowski, J. W. Kalat, and R. Nebes, "Reliability and Validity of Some Handedness Questionnaire Items," *Neuropsychologia* 12 (1974): 43–47.

8. Milan Dragovic, "Towards an Improved Measure of the Edinburgh Handedness Inventory: A One-Factor Congeneric Measurement Model Using Confirmatory Factor Analysis," *Laterality* 9 (2004): 411–19. S. Milenkovic and M. Dragovic, "Modification of the Edinburgh Handedness Inventory: A Replication Study," *Laterality* 18 (2013): 340–48.

9. L. Xin-tian, "The Distribution of Left and Right Handedness in Chinese

People," *Acta Psychologica Sinica* 3 (1983): 3, http://en.cnki.com.cn/Article_en /CJFDTOTAL-XLXB198303004.htm.

10. Oldfield, "The Assessment and Analysis of Handedness," 110.

11. Ibid., 99.

12. Xin-tian, "The Distribution of Left and Right Handedness in Chinese People."

13. "A Question of the Left Being Right-and Normal," *China Daily,* February 22, 2008.

14. Clare Porac, Laura Rees, and Terri Buller, "Switching Hands: A Place for Left Hand Use in a Right Hand World," in *Left-Handedness: Behavioral Implications and Anomalies,* ed. Stanley Coren, Advances in Psychology, no. 67 (Oxford, England: North-Holland, 1990), 259–90, 282; V. Llaurens, M. Raymond, and C. Faurie, "Why Are Some People Left-Handed?: An Evolutionary Perspective," *Philos Trans R Soc Lond B Biol Sci* 364 (2009): 881–94, 882.

15. Zverev, "Cultural and Environmental Pressure against Left-Hand Preference in Urban and Semi-Urban Malawi," 296.

16. H. I. Kushner, "Why Are There (Almost) No Left-Handers in China?," *Endeavour* 37 (June 2013): 71–81, 79; I. C. McManus et al., "Science in the Making: Right Hand, Left Hand, III: Estimating Historical Rates of Left-Handedness," *Laterality* 15 (2010): 186–208, 205.

17. Raymond and Pontier, "Is There Geographical Variation in Human Handedness?," 35–51, 45.

18. M. Peters, S. Reimers, and J. T. Manning, "Hand Preference for Writing and Associations with Selected Demographic and Behavioral Variables in 255,100 Subjects: The BBC Internet Study," *Brain Cogn* 62 (2006): 177–89.

19. Zverev, "Cultural and Environmental Pressure against Left-Hand Preference in Urban and Semi-Urban Malawi," 296.

20. Raymond and Pontier, "Is There Geographical Variation in Human Handedness?"; I. B. Perelle and L. Ehrman, "An International Study of Human Handedness: The Data," *Behav Genet* 24 (1994): 217–27.

21. I. C. McManus, "The Inheritance of Left-Handedness," *Ciba Found Symp* 162 (1991): 267–81, 257, 251–67; McManus, *Hypernotes to Right Hand, Left Hand: The Origins of Asymmetry in Brains, Bodies, Atoms, and Cultures,* 9:4, 2, www.righthandlefthand.com.

22. C. J. Wysocki and A. N. Gilbert, "National Geographic Smell Survey: Effects of Age Are Heterogenous," *Ann N Y Acad Sci* 561 (1989): 12–28; McManus, "The Inheritance of Left-Handedness," 251.

23. Herbert D. Chamberlain, "The Inheritance of Left-Handedness," *Journal of Heredity* 19 (1928): 557–59; K. R. Dronamraju, "Frequency of Left-Handedness among the Andhra Pradesh People," *Acta Genet Med Gemellol* (Roma) 24, no. 1–2 (1975): 161–62; M. Singh and M. P. Bryden, "The Factor Structure of Handedness in India," *Int J Neurosci* 74 (1994): 33–43.

24. Marietta Papadatou-Pastou et al., "Sex Differences in Left-Handedness: A Meta-Analysis of 144 Studies," *Psychol Bull* 134 (2008): 677–99, 677.

25. K. Kansaku, A. Yamaura, and S. Kitazawa, "Sex Differences in Lateralization Revealed in the Posterior Language Areas," *Cereb Cortex* 10 (2000): 866–72; B. A. Shaywitz et al., "Sex Differences in the Functional Organization of the Brain for Language," *Nature* 373 (1995): 607–9.

26. A. L. Foundas, C. M. Leonard, and B. Hanna-Pladdy, "Variability in the Anatomy of the Planum Temporale and Posterior Ascending Ramus: Do Right- and Left-Handers Differ?," *Brain Lang* 83 (2002): 403–24; J. J. Kulynych et al., "Gender Differences in the Normal Lateralization of the Supratemporal Cortex: MRI Surface-Rendering Morphometry of Heschl's Gyrus and the Planum Temporale," *Cereb Cortex* 4 (1994): 107–18.

27. Curtis Hardyck, Roy Goldman, and Lewis Petrinovich, "Handedness and Sex, Race, and Age," *Human Biology* 47, no. 3 (1975): 369–75.

28. D. A. Saunders, and A. L. Campbell, "Handedness Incidence in a Population of Black University Students," *Percept Mot Skills* 60, no. 2 (April 1985): 355–60, 59.

29. L. M. Lansky, H. Feinstein, and J. M. Peterson, "Demography of Handedness in Two Samples of Randomly Selected Adults (N = 2083)," *Neuropsychologia* 26, no. 3 (1988): 465–77, 467, 470.

30. S. Coren, *The Left-Hander Syndrome: The Causes and Consequences of Left-Handedness* (New York: Free Press, 1992), 50–51; S. Coren, "The Diminished Number of Older Left-Handers: Differential Mortality or Social-Historical Trend?," *Int J Neurosci* 75 (1994): 1–8.

31. M. De Agostini et al., "Environmental Influences in Hand Preference: An African Point of View," *Brain Cogn* 35 (1997): 151–67, 151, 157–63.

32. G. Dellatolas et al., "Age and Cohort Effects in Adult Handedness," *Neuropsychologia* 29 (1991): 255–61, 255.

33. Coren, *The Left-Hander Syndrome*, 206–21; D. F. Halpern and S. Coren, "Do Right-Handers Live Longer?," *Nature* 333 (1988): 213; D. F. Halpern and S. Coren, "Handedness and Life Span," *N Engl J Med* 324 (1991): 998.

34. I. C. McManus, *Right Hand, Left Hand: The Origins of Asymmetry in Brains, Bodies, Atoms, and Cultures* (London: Weidenfeld & Nicolson, 2002), 292–93; L. J. Harris, "Do Left-Handers Die Sooner than Right-Handers?: Commentary on Coren and Halpern's (1991) 'Left-Handedness: A Marker for Decreased Survival Fitness,'" *Psychol Bull* 114 (1993): 203–34, 235–47.

35. Evelyn Lee Teng et al., "Handedness in a Chinese Population: Biological, Social, and Pathological Factors," *Science* 193 (1976): 1148–50.

36. R. Hoosain, "Left Handedness and Handedness Switch amongst the Chinese," *Cortex* 26 (1990): 451–54.

37. M. De Agostini et al., "Environmental Influences in Hand Preference: An African Point of View," *Brain Cogn* 35 (1997): 151–67, 151, 157–63.

38. McManus, *Right Hand, Left Hand,* 151; T. Van Agtmael, S. M. Forrest, and R. Williamson, "Parametric and Non-Parametric Linkage Analysis of Several Candidate Regions for Genes for Human Handedness," *Eur J Hum Genet* 10 (2002): 623–30.

39. McManus, *Right Hand, Left Hand,* 149–51.

40. Ibid., 151; McManus, *Hypernotes to Right Hand, Left Hand,* 7:8, 3.

41. Melissa Roth, *The Left Stuff: How the Left-Handed Have Survived and Thrived* (Lanham, Md.: Rowman & Littlefield, 2005), 52.

42. Ibid.

43. Michael Peters, "Phenotype in Normal Left-Handers: An Understanding of Phenotype Is the Basis for Understanding Mechanism and Inheritance of Handedness," in *Left-Handedness: Behavioral Implications and Anomalies*, ed. Stanley

Coren, Advances in Psychology, no. 67 (Oxford, England: North-Holland, 1990), 167–92; Peters and Servos, "Performance of Subgroups of Left-Handers and Right-Handers"; McManus, *Right Hand, Left Hand*, 151.

44. Annett, "A Classification of Hand Preference by Association Analysis," 306; Roth, *The Left Stuff*, 54–55, 84.

45. McManus, *Right Hand, Left Hand*; McManus, *Hypernotes to Right Hand, Left Hand*, 7:8, 3.

46. M. L. Lalumière, R. Blanchard, and K. J. Zucker, "Sexual Orientation and Handedness in Men and Women: A Meta-Analysis," *Psychol Bull* 126 (2000): 575–92, 579.

47. McManus, *Hypernotes to Right Hand, Left Hand*, 7:13, 5.

48. Roth, *The Left Stuff*, 54; M. Annett, "Schizophrenia and Autism Considered as the Products of an Agnosic Right Shift Gene," *Cogn Neuropsychiatry* 2 (1997): 195–214.

49. John L. Dawson, "Temne-Arunta Hand-Eye Dominance and Cognitive Style," *International Journal of Psychology* 7 (1972): 219–33, 219, 222.

50. Ibid.

51. Zverev, "Cultural and Environmental Pressure against Left-Hand Preference in Urban and Semi-Urban Malawi," 297.

52. M. P. Bryden, A. Ardila, and O. Ardila, "Handedness in Native Amazonians," *Neuropsychologia* 31 (1993): 301–8.

53. Ibid., 306.

54. Dronamraju, "Frequency of Left-Handedness among the Andhra Pradesh People."

55. The Association of Left Handers, "The Indian Perspective" (2010), http://www.lefthanders.org/lefties/.

56. C. J. Brackenridge, "Secular Variation in Handedness over Ninety Years," *Neuropsychologia* 19 (1981): 459–62.

57. Leen J. Beukelaar and Pieter M. Kroonenberg, "Changes over Time in the Relationship between Hand Preference and Writing Hand among Left-Handers," *Neuropsychologia* 24 (1986): 301–3.

58. J. J. Terrell, "Let Left-Handedness Alone," *Illustrated World* 27 (1917); Harvey Ernest Jordan, "The Crime against Left-Handedness," *Good Health* 57 (1922): 378–83; L. J. Harris, "Left-Handedness: Early Theories, Facts, and Fancies," in *Neuropsychology of Left-Handedness*, ed. Jeannine Herron, Perspectives in Neurolinguistics and Psycholinguistics (New York: Academic Press, 1980): 63–65, 216–18.

59. Wysocki and Gilbert, "National Geographic Smell Survey."

60. Harris suggested that "a cross-generational, or 'secular trend,' analysis" might provide even clearer evidence for this association. Assuming that "the prevalence of left-handedness in a particular age cohort . . . reflects cultural practices pertaining to handedness," cohorts could then be compared over time. L. J. Harris, "Cultural Influences on Handedness: Historical and Contemporary Theory and Evidence," in *Left-Handedness: Behavioral Implications and Anomalies*, ed. Stanley Coren, Advances in Psychology, no. 67 (Oxford, England: North-Holland, 1990), 195–258, 217–18.

61. A. N. Gilbert and C. J. Wysocki, "Hand Preference and Age in the United States," *Neuropsychologia* 30 (1992): 601–8, 603, 605.

62. McManus et al., "Science in the Making," 186, 205.

63. Halpern and Coren, "Do Right-Handers Live Longer?"

64. McManus, "The Inheritance of Left-Handedness"; McManus, *Right Hand, Left Hand.*

65. De Agostini et al., "Environmental Influences in Hand Preference," 152; G. Dellatolas et al., "Mesure De La Préférence Manuelle Par Autoquestionnaire Dans La Population Française Adulte," *Revue de Psychologie Appliquée* 38 (1988): 117–35.

66. De Agostini et al., "Environmental Influences in Hand Preference," 152.

67. McManus et al., "Science in the Making," 186, 205.

68. H. I. Kushner, "Retraining Left-Handers and the Aetiology of Stuttering: The Rise and Fall of an Intriguing Theory," *Laterality* 17, no. 6 (2012): 673–93.

69. Oldfield, "The Assessment and Analysis of Handedness," 110.

70. Corballis, "The Genetics and Evolution of Handedness," 714.

CHAPTER 4. AMBIGUOUS ATTITUDES

Epigraphs: Robert Williams (pseudonym), email to Howard Kushner, July 18, 2016; Lao Tse, *The Tao Te Ching: Or the Tao and Its Characteristics,* vol. 39, trans. James Legge (1891), Sacred Books of the East, http://www.gutenberg.org/etext/23974, retrieved Aug. 13, 2010: 31.1; Dudley Kidd, *Savage Childhood: A Study of Kafir Children* (London: Adam and Charles Black, 1906), 296.

1. Dudley Kidd, *Savage Childhood: A Study of Kafir Children* (London: Adam and Charles Black, 1906), 296.

2. Heinz A. Wieschhoff, "Concepts of Right and Left in African Cultures," *Journal of the American Oriental Society* 58 (1938): 202–17, 216–17; John L. Dawson, "Temne-Arunta Hand-Eye Dominance and Cognitive Style," *International Journal of Psychology* 7 (1972): 219–33.

3. Howard I. Kushner, "What We Do (and Don't) Know about Left-Handedness," March 21, 2016, *Late Night Live,* ABC Australia, www.abc.net.au/radionational /programs/latenightlive/what-we-do-and-dont-know-about-left-handedness /7254986.

4. Michael Hackh, email to Howard Kishner, August 1, 2016, http:// linkshaenderforum.org/forum/index.php.

5. Ibid.

6. Samuel M. Randolph, *Hidden Handedness: The Emerging Story of Handedness Reversals* (n.p: Virtualbookworm.com Publishing, 2007).

7. S. T. Orton, *Reading, Writing and Speech Problems in Children* (New York: W. W. Norton & Co., 1937), 50; also see L. J. Harris, "Cultural Influences on Handedness: Historical and Contemporary Theory and Evidence," in *Left-Handedness: Behavioral Implications and Anomalies,* ed. Stanley Coren, Advances in Psychology, no. 67 (Oxford, England: North-Holland, 1990), 195–258, 197, 199–200; P-M. Bertrand, *Histoire de gauchers: Des gens a l'enfers* (Paris: Éditions Imago, 2001), 91–102.

8. I. C. McManus, *Right Hand, Left Hand: The Origins of Asymmetry in Brains, Bodies, Atoms, and Cultures* (London: Weidenfeld & Nicolson, 2002), 268.

9. Ibid.

10. Mark Logue and Peter Conradi, *The King's Speech: How One Man Saved the British Monarchy* (New York: Sterling Publishing Co., 2010).

11. Williams, email to Howard Kushner, July 18, 2016.

12. S. Coren, *The Left-Hander Syndrome: The Causes and Consequences of Left-Handedness* (New York: Free Press, 1992), 65–66.

13. Les Gauchers, "L'école Des Années 50 et Les Gauchers," *Gauchers, Journale Internationale des Gauchers* no. 13 (June 2014), www.lesgauchers.com/chroniques -pour-gaucher/gaucher-1-ecole-des-annees-50.

14. Domitille Arrivet, "Les Entreprises Multiplient Les Produits Conçus Pour Les Gauchers," *Le Figaro Economie* (August 13, 2016).

15. C. Porac, S. Coren, and A. Searleman, "Environmental Factors in Hand Preference Formation: Evidence from Attempts to Switch the Preferred Hand," *Behav Genet* 16 (1986): 251–61.

16. Coren, *The Left-Hander Syndrome*, 68.

17. Natalie Jacewicz, "Right-Hand Bias Is Everywhere," *The Atlantic*, March 8, 2016.

18. M. A. Payne, "Impact of Cultural Pressures on Self-Reports of Actual and Approved Hand Use," *Neuropsychologia* 25 (1987): 247–58, 254–58.

19. Bertrand provides an extensive historical examination of the contrast between "the good hand and the bad hand," whereby right hand is synonymous with the positive and sacred and the left serves as a metaphor for negative and bad. Bertrand, *Histoire de gauchers*, 27–39; Wulf Schiefenhövel, "Biased Semantics for Right and Left in 50 Indo-European and Non-Indo-European Languages," *Ann NY Acad Sci* 1288 (2013): 135–52.

20. McManus, *Right Hand, Left Hand*, 60–65.

21. Coren, *The Left-Hander Syndrome*, 1–3; also see M. Barsley, *The Other Hand: An Investigation into the Sinister History of Left-Handedness* (Portland, Ore.: Hawthorne Books, 1967).

22. Walter H. Stapleton, "The Terms for 'Right Hand' and 'Left Hand' in the Bantu Languages," *Journal of the Royal African Society* 4, no. 16 (1905): 431–33; A. Werner, "Note on the Terms Used for 'Right Hand' and 'Left Hand' in the Bantu Languages," *Journal of the Royal African Society* 4, no. 13 (1904): 112–16.

23. Robert Hertz, "La Prééminence de la Main Doite: Étude sur la Polarité Religieuse," *Revue Philosophique* LXVIII (1909): 553–80; Hertz, *Death and the Right Hand*, ed. and intro. E. E. Evans-Pritchard, trans. Rodney Needham and Claudia Needham (New York: Routledge, 1960; reprinted 2004), 98.

24. Al-Tabari quoted in J. Chelhod, "A Contribution to the Problem of the Pre-Eminance of the Right Based on Arabic Evidence," in *Right & Left: Essays on Dual Symbolic Classification*, ed. Rodney Needham (Chicago: University of Chicago Press, 1973): 239–62, 240.

25. Ibid.

26. Wieschhoff, "Concepts of Right and Left in African Cultures," 202–17, 216.

27. Y. P. Zverev, "Cultural and Environmental Pressure against Left-Hand Preference in Urban and Semi-Urban Malawi," *Brain and Cognition* 60 (2006): 295–30, 297.

28. Ibid., 299–300.

29. Melissa Roth, *The Left Stuff: How the Left-Handed Have Survived and Thrived* (Lanham, Md.: Rowman & Littlefield, 2005), 27–28.

30. Bertrand, *Histoire de gauchers*, 65–67.

31. Plato, "Laws (First Printed 1926)," in *Laws*, 2 vols., ed. and trans. R. G. Bury (Cambridge, Mass.: Harvard/London/Heinemann, 1984), 2: 794c–795c.

32. Daniel G. Brinton, "Left-Handedness in North American Aboriginal Art," *American Anthropologist* 9 (1896): 175–81, 176–77.

33. Lao Tse, *The Tao Te Ching*, 31.1.

34. Marcel Granet, "Right and Left in China," in Needham, *Right & Left*, 43–58, 44.

35. Bertrand, *Histoire de gauchers*, 172–94, quotation 173.

36. Schiefenhövel, "Biased Semantics for Right and Left in 50 Indo-European and Non-Indo-European Languages," 146–47, 135.

37. Wieschhoff, "Concepts of Right and Left in African Cultures," 217.

38. Brinton, "Left-Handedness in North American Aboriginal Art," 176.

39. Barsley, *The Other Hand*, 30.

40. John Vernon, *The Last Canyon* (Boston: Houghton Mifflin, 2001), 50.

41. Albert C. Kruyt, "Right and Left in Central Celebes," in Needham, *Right & Left*, 74–91, 74–75.

42. Ibid., 80, 83.

43. Hertz, *Death and the Right Hand*.

44. A. N. Gilbert and C. J. Wysocki, "Hand Preference and Age in the United States," *Neuropsychologia* 30 (1992): 601–8, 606.

45. L. Xin-tian, "The Distribution of Left and Right Handedness in Chinese People," *Acta Psychologica Sinica* 3 (1983), http://en.cnki.com.cn/Article_en /CJFDTOTAL-XLXB198303004.htm.

46. "A Question of the Left Being Right—and Normal," *China Daily*, February 22, 2008.

47. Evelyn Lee Teng et al., "Handedness in a Chinese Population: Biological, Social, and Pathological Factors," *Science* 193 (1976): 1148–50.

48. Ching-Chang Hung et al., "A Study on Handedness and Cerebral Speech Dominance in Right-Handed Chinese," *Journal of Neurolinguistics* 1 (1985): 143–63, 143.

49. C. Quinan, "The Principal Sinistral Types: An Experimental Study Particularly as Regards Their Relation to the So-Called Constitutional Psychopathic States," *Archives of Neurology & Psychiatry* (Chicago) 24 (1930): 35–47, 40–41.

50. Teng et al., "Handedness in a Chinese Population," 1149.

51. M. Singh and M. P. Bryden, "The Factor Structure of Handedness in India," *Int J Neurosci* 74 (1994): 33–43.

52. M. Singh, M. Manjary, and G. Dellatolas, "Lateral Preferences among Indian School Children," *Cortex* 37, no. 2 (2001): 231–41, 233–35.

53. Ibid., 235–37.

54. M. De Agostini et al., "Environmental Influences in Hand Preference: An African Point of View," *Brain Cogn* 35 (1997): 151–67, 151, 162–65.

55. Ibid., 152.

56. Ibid., 151, 162–65.

57. "A Question of the Left Being Right—and Normal."

58. Ibid.

59. Lydie Mepham, "Thoughts on Handedness in China," email to author, July 28, 2011.

60. K. R. Dronamraju, "Frequency of Left-Handedness among the Andhra Pradesh People," *Acta Genet Med Gemellol* (Roma) 24, no. 1–2 (1975): 161–62.

61. Zverev, "Cultural and Environmental Pressure against Left-Hand Preference in Urban and Semi-Urban Malawi," 296.

62. Clare Porac, Laura Rees, and Terri Buller, "Switching Hands: A Place for Left Hand Use in a Right Hand World," in Coren, *Left-Handedness*, 259–90, 285.

63. Harris, "Cultural Influences on Handedness," 196.

64. L. J. Harris, "What to Do About Your Child's Handedness?: Advice from Five Eighteenth-Century Authors, and Some Questions for Today," *Laterality* 8 (2003): 99–120.

65. Harris, "Cultural Influences on Handedness," 216–17; Gilbert and Wysocki, "Hand Preference and Age in the United States," 603–5; I. C. McManus et al., "Science in the Making: Right Hand, Left Hand, III: Estimating Historical Rates of Left-Handedness," *Laterality* 15 (2010): 186–208, 186, 205.

66. Hertz, "La Prééminence de la Main Doite"; H. I. Kushner, "Deficit or Creativity: Cesare Lombroso, Robert Hertz, and the Meanings of Left-Handedness," *Laterality* 18, no. 4 (2013): 416–36.

67. Kushner, "Deficit or Creativity."

Chapter 5. Changing Hands, Tying Tongues

Epigraphs: Harvey Ernest Jordan, "The Crime against Left-Handedness," *Good Health* 57 (1922): 373–83, 382–83; Abram Blau, *The Master Hand: A Study of the Origin and Meaning of Left and Right Sidedness and Its Relation to Personality and Language* (New York: American Orthopsychiatric Association, 1946), 113, 185.

1. Marc Shell, *Stutter* (Cambridge, Mass.: Harvard University Press, 2005), 7–20.

2. H. I. Kushner, "Cesare Lombroso and the Pathology of Left-Handedness," *Lancet* 377 (2011): 118–19.

3. H. I. Kushner, "Retraining Left-Handers and the Aetiology of Stuttering: The Rise and Fall of an Intriguing Theory," *Laterality* 17, no. 6 (2012): 673–93.

4. William Elder, "The Left-Handed Child," *British Medical Journal* 2 (1924): 80–81.

5. H. Drinkwater, "The Left-Handed Child (Letter)," *British Medical Journal* 1, no. 3312 (1924): 1113. For a comprehensive discussion, see P-M. Bertrand, *Histoire de gauchers: Des gens a l'enfers* (Paris: Éditions Imago, 2001), 75–85.

6. Mark Logue and Peter Conradi, *The King's Speech: How One Man Saved the British Monarchy* (New York: Sterling Publishing Co., 2010).

7. S. J. W. Wheeler-Bennett, *King George VI: His Life and Reign* (New York: St. Martin's Press, 1958), quoted in M. Barsley, *The Other Hand: An Investigation into the Sinister History of Left-Handedness* (Portland, Ore.: Hawthorne Books, 1967), 169; also see Logue and Conradi, *The King's Speech*, 51.

8. L. J. Harris, "Cultural Influences on Handedness: Historical and Contemporary Theory and Evidence," in *Left-Handedness: Behavioral Implications and Anomalies*, ed. Stanley Coren, Advances in Psychology, no. 67 (Oxford, England: North-Holland, 1990), 195–258, 213.

9. Smith's survey of Trenton State School for Girls, the Hallowell State School for Girls, and the Shirley Industrial School for Boys revealed that "15% of the boys and 6.5% of the girls are left-handed." Moreover, she found that in institutions

for the feebleminded in Maine and Rhode Island, 11 percent of the girls and 8.5 percent of the boys were left-handed. Laura G. Smith, "A Brief Survey of Right and Left-Handedness," *Pedagogical Seminary* 24 (1917): 19–35, 32.

10. Ibid.

11. J. J. Terrell, "Let Left-Handedness Alone," *Illustrated World* 27 (1917): 190–92.

12. Jordan, "The Crime against Left-Handedness," 379, 382–83.

13. There was no single definition or agreed practice for what constituted eugenics. See Frank Dikötter, "Race Culture: Recent Perspectives on the History of Eugenics," *Am Hist Rev* 103 (1998): 467–78.

14. Harris, "Cultural Influences on Handedness," 213.

15. P. B. Ballard, "Sinistrality and Speech," *Journal of Experimental Pedagogy and Training College Record* 1 (1911, 1912): 298–310, 308. The first study was based on a questionnaire sent to 13,189 South London schoolteachers. Of the 545 children who were left-handed, 399 had been retrained to write with the right hand (299–300). Ballard labeled these children "dextro-sinistrals" (*dextro* for right, *sinistral* for left). Only 1.1 percent of the left-handed children who were not switched stuttered, compared to 4.3 percent of the forced right-handers. Ballard's second study found that while only 1.6 percent of right-handers were stutterers, 20 percent of the dextro-sinistrals stuttered. Ballard's third study of 11,939 schoolchildren was even more persuasive. Of the 51 left-handers who had been allowed to use their left hands, none stuttered.

16. Lewis Madison Terman, *The Hygiene of the School Child* (Cambridge, Mass.: Riverside Press, 1914), 345–46.

17. Elder, "The Left-Handed Child," 80–81.

18. D. K. Mohlman, "A Preliminary Study of the Problems in the Training of the Non-Preferred Hand," *Journal of Educational Psychology* 14 (1923): 215–30, 216.

19. Ira S. Wile, "The Relation of Left-Handedness to Behavior Disorders," *American Journal of Orthopsychiatry* 2 (1932): 44–57, 45.

20. Ira S. Wile, *Handedness: Right and Left* (Oxford, England: Lothrop, 1934), 348.

21. Wile, "The Relation of Left-Handedness to Behavior Disorders," 49–55.

22. Wile, *Handedness: Right and Left*, 351.

23. S. T. Orton and L. E. Travis, "Studies in Stuttering: Studies of Action Currents in Stutterers," *Archives of Neurology & Psychiatry* (Chicago) 21 (1929): 61–68.

24. S. T. Orton, "Studies in Stuttering: Introduction," *Archives of Neurology & Psychiatry* (Chicago) 18 (1927): 671–72.

25. Dyslexia was first identified in 1887 by the German ophthalmologist Rudolph Berlin. I thank Lauren J. Harris for pointing this out.

26. S. T. Orton, *Reading, Writing, and Speech Problems in Children* (New York: W. W. Norton & Co., 1937), 200.

27. Orton and Travis, "Studies in Stuttering," 27.

28. Orton, *Reading, Writing, and Speech Problems in Children*, 200.

29. Lee Edward Travis, *Speech Pathology: A Dynamic Neurological Treatment of Normal Speech and Speech Deviations* (New York: D. Appleton & Co., 1931), 185.

30. Logue and Conradi, *The King's Speech*, 132.

31. Wendell Johnson and Lucile Duke, "Changes in Handedness Associated with Onset or Disappearance of Stuttering: Sixteen Cases," *Journal of Experimental Education* 4 (September 1935): 112–32.

32. Travis, *Speech Pathology*, 56, 179–80.

33. Ibid., 190–91. Travis believed that other interventions, including psychotherapy, were also useful additions for treating stutterers: "A main goal to be achieved in the education of the stutterer's attitude toward his trouble is his impersonal evaluation of disability. He must learn to objectify it, and in some cases accept it." Ibid., 184; also see Marjorie Van de Water, "Why Stutterers Stutter," *Science News-Letter* 21, no. 570 (1932): 162–68.

34. B. Bobrick, *Knotted Tongues: Stuttering in History and the Quest for a Cure* (Simon & Schuster, 1995), 129.

35. Johnson and Duke, "Changes in Handedness Associated with Onset or Disappearance of Stuttering," 114–15.

36. Ibid., 112, 131–32.

37. To test his hypothesis, in 1939, Johnson recruited a graduate student, Mary Tudor, to conduct a study of 22 children at the Davenport, Iowa, Soldiers and Sailors Orphans' Home. For the next six months 10 of the children were subjected to extremely negative reactions to their speech, while the rest received exceptionally positive responses. Students experiencing negative reactions developed severe, lifelong psychological emotional deficits; the controls were free of all pathology. When colleagues learned of the "experiment," they were generally horrified and labeled it "The Monster Study." In 2001, the University of Iowa apologized for the study and in 2007, six of the orphan children were awarded approximately $1 million by the State of Iowa. Mary Tudor, "An Experimental Study of the Effect of Evaluative Labeling of Speech Fluency" (MA thesis, State University of Iowa, 1939), http://ir.uiowa.edu/cgi/viewcontent.cgi?article=6264&context=etd; Gretchen Reynolds, "The Stuttering Doctor's 'Monster Study,'" *New York Times*, March 16, 2003, http://www.nytimes.com/2003/03/16/magazine/the-stuttering-doctor-s -monster-study.html; Franklin H. Silverman, "The 'Monster' Study," *Journal of Fluency Disorders* 13 (June 1, 1988): 225–31.

38. Orton, *Reading, Writing, and Speech Problems in Children*, 194–95.

39. B. Bryngelson, "A Method of Stuttering," *Journal of Abnormal and Social Psychology* 30 (1935): 194–98, 194.

40. B. Bryngelson and T. B. Clark, "Left-Handedness and Stuttering," *Journal of Heredity* 24 (1933): 387–90, 387.

41. B. Bryngelson, "A Study of Laterality of Stutterers and Normal Speakers," *Journal of Social Psychology* 11 (1940): 151–55, 152.

42. Ibid., 153.

43. W. R. Brain, "Speech and Handedness," *Lancet* 249 (1945): 837–41, 837.

44. Beaufort Sims Parson, *Left-Handedness: A New Interpretation: Foreword by Harvey E. Jordan, A.M., Ph.D.* (New York: Macmillan, 1924), 99.

45. A study of more than 1,100 elementary students in Nicosia, Cyprus, found that none of the stutterers had been switched to right-handers. Kypros Chrysanthis, "Stammering and Handedness," *Lancet* 249, no. 6442 (February 15, 1947): 270–71.

46. K. C. Garrison, "Problems Related to Left-Handedness," *Peabody Journal of Education* 15 (1938): 325–32, 328.

47. E. Shorter, *A History of Psychiatry* (New York: J. Wiley, 1997), 145–89.

48. H. I. Kushner, *A Cursing Brain?: The Histories of Tourette Syndrome* (Cambridge, Mass.: Harvard University Press, 1999), 172.

49. Ibid., 99–118.

50. Isador H. Coriat, "The Nature and Analytical Treatment of Stuttering," *Proc Am Speech Correction Association* 1 (1931): 151–56, 154–55.

51. Kushner, *A Cursing Brain?*, 99–118.

52. Else Heilpern, "A Case of Stuttering," *Psychoanalytic Quarterly* 10 (1941): 94–115, 94–95.

53. O. Fenichel, *Hysterien Und Zwangsneurosen. Vienna* (Vienna: Int. Psychoanalytischer Verlag, 1931).

54. Heilpern, "A Case of Stuttering," 94–95.

55. O. Fenichel, *The Psychoanalytic Theory of Neurosis* (New York: W. W. Norton & Co., 1945): 288–90.

56. Blau, *The Master Hand*.

57. In contrast, in 1947 a committee of the French Education Ministry issued a report condemning forced switching of left-handers to writing with their right hands. Although the report never became law, it does suggest that retraining was controversial in postwar France. Domitille Arrivet, "Les Entreprises Multiplient Les Produits Conçus Pour Les Gauchers," *Le Figaro Economie* (August 13, 2016). Also see Les Gauchers, "L'école Des Années 50 et Les Gauchers," *Gauchers, Journale Internationale des Gauchers* no. 13 (June 2014), www.lesgauchers.com/chroniques -pour-gaucher/gaucher-1-ecole-des-annees-50.

58. Gertrude Hildreth, "The Development and Training of Hand Dominance, IV: Developmental Problems Associated with Handedness," *Pedagogical Seminary and Journal of Genetic Psychology* 76 (1950): 101–54, 101.

59. Kenneth L. Martin, "Handedness: A Review of the Literature on the History, Development and Research of Laterality Preference," *Journal of Educational Research* 45 (1952): 527–33, 531.

60. M.A. Stanford, "It's No Fun to Be a Southpaw," *Parents' Magazine* 18, November 1943, 24, 76.

61. Margaret M. Clark, *Left-Handedness: Laterality Characteristics and Their Educational Implications* (London: University of London Press, 1957), 50.

62. Oliver Bloodstein, *Stuttering for Professional Workers* (Chicago: National Society for Crippled Children and Adults, 1959), 35, quoted in Charles Van Riper, *The Nature of Stuttering* (Prentice-Hall: Englewood Cliffs, N.J.: 1971), 357.

63. R. K. Jones, "Observations on Stammering after Localized Cerebral Injury," *Journal of Neurology, Neurosurgery & Psychiatry* 29 (1966): 192–95, 195.

64. Michael C. Corballis and Ivan L. Beale, *The Ambivalent Mind: The Neuropsychology of Left & Right* (Chicago: Nelson Hall, 1983), 229, 234–35.

65. Harris, "Cultural Influences on Handedness," 226–27.

66. C. J. Brackenridge, "Secular Variation in Handedness over Ninety Years," *Neuropsychologia* 19 (1981): 459–62, 459.

67. M. C. Corballis, *The Lopsided Ape: Evolution of the Generative Mind* (New York: Oxford University Press, 1991), 89.

68. C. Richard Dean and Robert A. Brown, "A More Recent Look at the Prevalence of Stuttering in the United States," *Journal of Fluency Disorders* 2 (1977): 157–66.

69. Corballis, *The Lopsided Ape*, 201.

70. A. L. Foundas et al., "Anomalous Anatomy of Speech-Language Areas in Adults with Persistent Developmental Stuttering," *Neurology* 57, no. 2 (2001): 207–15, 207.

71. A. L. Foundas et al., "Aberrant Auditory Processing and Atypical Planum Temporale in Developmental Stuttering," *Neurology* 63, no. 9 (2004): 1640–46; A. L. Foundas et al., "Atypical Cerebral Laterality in Adults with Persistent Developmental Stuttering," *Neurology* 61, no. 10 (2003): 1378–85.

72. C. Buchel and M. Sommer, "What Causes Stuttering?," *PLoS Biol* 2 (2004): E46.

73. P. A. Alm et al., "Hemispheric Lateralization of Motor Thresholds in Relation to Stuttering," *PLoS One* 8, no. 10 (2013): e76824.

74. Buchel and Sommer, "What Causes Stuttering?," 163.

75. L. Jancke, J. Hanggi, and H. Steinmetz, "Morphological Brain Differences between Adult Stutterers and Non-Stutterers," *BMC Neurol* 4 (2004): 23.

76. Van Riper, *The Nature of Stuttering,* 356.

77. M. C. Corballis, "Left Brain, Right Brain: Facts and Fantasies," *PLoS Biol* 12, no. 1 (2014): e1001767.

78. H. Kathard, "Issues of Culture and Stuttering: A South African Perspective," 1998, paper published for the First International Stuttering Awareness Day online conference, mnsu.edu/dept/comdis/isad/papers/kathard.html.

79. A. K. Mishra and Ruchika Gupta, "Disability Index: A Measure of Deprivation among Disabled," *Economic and Political Weekly* 41 (2006): 4026–29; Sophie Mitra and Usha Sambamoorthi, "Disability Estimates in India: What the Census and Nss Tell Us," *Economic and Political Weekly* 41, no. 38 (2006): 4022–26; also see Shoba Srinath et al., "Epidemiological Study of Child & Adolescent Psychiatric Disorders in Urban & Rural Areas of Bangalore, India," *Indian J Med Res* 122 (July 2005): 67–79, 72.

80. J. X. Ming et al., "Public Awareness of Stuttering in Shanghai, China," *Logopedics, Phoniatrics, Vocology* 26 (2001): 145–50.

81. H. Mohammadi et al., "Late Recovery from Stuttering: The Role of Hand Dominancy, Fine Motor and Inhibition Control," *Iran J Psychiatry* 11 (2016): 51–58.

CHAPTER 6. FROM GENES TO POPULATIONS

Epigraphs: John B. Watson, "What the Nursery Has to Say about Instincts," *Ped Sem* 32 (1925): 321–22; Arnold Gesell and Louise Bates Ames, "Early Evidences of Individuality in the Human Infant," *Scientific Monthly* 45 (1937): 217–25, 225.

1. H. I. Kushner, "Deficit or Creativity: Cesare Lombroso, Robert Hertz, and the Meanings of Left-Handedness," *Laterality* 18, no. 4 (2013): 416–36.

2. Wendell Johnson and Lucile Duke, "Changes in Handedness Associated with Onset or Disappearance of Stuttering: Sixteen Cases," *Journal of Experimental Education* 4 (1935): 112–32; H. I. Kushner, "Retraining Left-Handers and the Aetiology of Stuttering: The Rise and Fall of an Intriguing Theory," *Laterality* 17, no. 6 (2012): 673–93.

3. S. M. Schaafsma et al., "The Relation between Handedness Indices and Reproductive Success in a Non-Industrial Society," *PLoS One* 8, no. 5 (2013): e63114.

4. George M. Gould, *Righthandedness and Lefthandedness, with Chapters Treating of the Writing Posture, the Rule of the Road* (Philadelphia: J. B. Lippincott, 1908).

5. George Milbry Gould and Walter Lytle Pyle, *Anomalies and Curiosities of Medicine: Being an Encyclopedic Collection of Rare and Extraordinary Cases, and of the Most Striking Instances of Abnormality in All Branches of Medicine and Surgery, Derived from and Exhaustive Research of Medical Literature Its Origin to the Present*

Day, Abstracted, Classified, Annotated and Indexed (Philadelphia: W. B. Saunders, 1896).

6. Gould, *Righthandedness and Lefthandedness*, 44.

7. "The Riddle of Righthandedness Made Clear," *New York Times*, May 31, 1908.

8. Haywood Broun, "Let the Southpaw Alone," *Collier's, the National Weekly*, April 3, 1920, 22.

9. "Interesting Items from the World's Press," *The Daily News of Perth* (Australia), January 18, 1908, 2.

10. Harvey Ernest Jordan, "The Inheritance of Left-Handedness," *American Breeders' Magazine* 2 (1911): 19–21.

11. Frank Dikötter, "Race Culture: Recent Perspectives on the History of Eugenics," *Am Hist Rev* 103 (1998): 467–78.

12. Harvey Ernest Jordan, "The Crime against Left-Handedness," *Good Health* 57 (1922): 378–83, 381–82.

13. Beaufort Sims Parson, *Left-Handedness: A New Interpretation. Foreword by Harvey E. Jordan, A.M., Ph.D.* (New York: Macmillan, 1924), 46, 69, 127, quoting H. E. Jordan, "Hereditary Lefthandedness," *Journal of Genetics* 4 (1914): 67–77, and Jordan, "The Inheritance of Left-Handedness," 77.

14. Parson, *Left-Handedness*, 69.

15. Ibid., 68–69.

16. For instance, see W. C. Reavis, "Review: A New Interpretation of Left-Handedness," *School Review* 33 (April 1925): 312–13.

17. June E. Downey, "Review of Parson *Lefthandedness: A New Interpretation*," *Journal of Abnormal and Social Psychology* 20 (June 4, 1926): 452–55.

18. "Review of Parson, *Left-Handedness: A New Interpretation*," *British Medical Journal* 2 (3379) (October 3, 1925): 620–21.

19. Stevenson Smith, "Review: *Left-Handedness* by Beaufort Sims Parson," *Science* 63 (April 9, 1926): 383–84. Smith was particularly critical of Parson's manuscope, which he found "inadequately described" and mechanically primitive. Moreover, Parson had produced no evidence that the manuscope met any test of its reliability (384).

20. R. E. Rosch, D. V. Bishop, and N. A. Badcock, "Lateralised Visual Attention Is Unrelated to Language Lateralisation, and Not Influenced by Task Difficulty—A Functional Transcranial Doppler Study," *Neuropsychologia* 50 (2012): 810–14, 813–14.

21. W. M. Brandler et al., "Common Variants in Left/Right Asymmetry Genes and Pathways Are Associated with Relative Hand Skill," *PLoS Genet* 9 (2013): e1003751; W. M. Brandler and S. Paracchini, "The Genetic Relationship between Handedness and Neurodevelopmental Disorders," *Trends Mol Med* 20 (February 2014): 83–90.

22. Honor Whiteman, "Left and Right-Handed Genes Discovered," *Medical News Today* (September 16, 2013), www.medicalnewstoday.com/articles/266093.php.

23. J. A. Armour, A. Davison, and I. C. McManus, "Genome-Wide Association Study of Handedness Excludes Simple Genetic Models," *Heredity* (Edinburgh) 112 (March 2014): 221–25.

24. Andrey Giljov et al., "Parallel Emergence of True Handedness in the Evolution of Marsupials and Placentals," *Current Biology* 25 (2015): 1878–84; C. Spiezio, B. Regaiolli, and G. Vallortigara, "Motor and Postural Asymmetries in Marsupials:

Forelimb Preferences in the Red-Necked Wallaby (Macropus Rufogriseus)," *Behav Processes* 128 (2016): 119–25.

25. Emily Willingham, "Kangaroos Join Ranks of Animals Reported to Have Autism," *Forbes*, June 21, 2015, www.forbes.com/sites/emilywillingham/2015/06/21/kangaroos-join-ranks-of-animals-reported-to-have-autism/.

26. Jordan, "The Crime against Left-Handedness," 381–82.

27. Ibid., 23–24, 27–28.

28. Francis Ramaley, "Inheritance of Left-Handedness," *American Naturalist* 47, no. 564 (1913): 730–38, 735, 738.

29. Herbert D. Chamberlain, "The Inheritance of Left-Handedness," *Journal of Heredity* 19 (1928): 557–59.

30. Ibid., 557.

31. R. A. Fisher, "The Goodness of Fit of Regression Formulae, and the Distribution of Regression Coefficients," *Journal of the Royal Statistical Society* 85 (1922): 597–612.

32. Even though Fisher's 1922 method was theoretically available to Chamberlain, it was, according to statistician John Aldrich, extremely difficult to use and it would have been exceedingly labor intensive to employ. Before the 1970s and the development of statistical software packages, regression tests took many hours, even days, to run. John Aldrich, "Fisher and Regression," *Statist Sci* 24 (2005): 401–17.

33. Gesell and Ames, "Early Evidences of Individuality in the Human Infant," 221–22, 225.

34. K. C. Garrison, "Problems Related to Left-Handedness," *Peabody Journal of Education* 15 (1938): 325–32, 325–27.

35. John Broadus Watson, *Psychology, from the Standpoint of a Behaviorist* (Philadelphia: J. B. Lippincott Company, 1919), 241.

36. Watson, "What the Nursery Has to Say about Instincts," 321–22.

37. Abram Blau, *The Master Hand: A Study of the Origin and Meaning of Left and Right Sidedness and Its Relation to Personality and Language* (New York: American Orthopsychiatric Association, 1946), 91.

38. Ibid., quotation on 5; also see 19, 21–22, 93.

39. Minnie Giesecke, "The Genesis of Hand Preference," *Monographs of the Society for Research in Child Development*, vol. 1 (London: Blackwell Publishing on behalf of the Society for Research in Child Development, 1936), 1–102; W. Ludwig, *Reohts-Links-Problem Im Tierreloh Und Beim Mensohen* (Berlin: Julius Springer, 1932).

40. Giesecke, "The Genesis of Hand Preference," 5, 80.

41. Ludwig, *Reohts-Links-Problem Im Tierreloh Und Beim Mensohen*, 353.

42. Giesecke, "The Genesis of Hand Preference," 80.

43. Margaret M. Clark, *Left-Handedness: Laterality Characteristics and Their Educational Implications* (London: University of London Press, 1957), 17–18.

44. Jerre Levy and Thomas Nagylaki, "A Model for the Genetics of Handedness," *Genetics* 72 (1972): 117–28.

45. Patrick T. W. Hudson, "The Genetics of Handedness—A Reply to Levy and Nagylaki," *Neuropsychologia* 13 (1975): 331–39.

46. Marian Annett, "The Genetics of Handedness," *Trends in Neuroscience* 4 (1981): 256–58, 257.

47. G. C. Ashton, "Handedness: An Alternative Hypothesis," *Behav Genet* 12 (1982): 125–47, 125–26.

48. P. Bakan, G. Dibb, and P. Reed, "Handedness and Birth Stress," *Neuropsychologia* 11 (1973): 363–66.

49. S. Coren and C. Porac, "Birth Factors and Laterality: Effects of Birth Order, Parental Age, and Birth Stress on Four Indices of Lateral Preference," *Behav Genet* 10 (1980): 123–38.

50. Ashton, "Handedness: An Alternative Hypothesis," 144–46.

51. N. Geschwind, "Biological Associations of Left-Handedness," *Annals of Dyslexia* 33 (1983): 29–40; N. Geschwind and P. O. Behan, "Left-Handedness: Association with Immune Disease, Migraine, and Developmental Learning Disorder," *Proc Natl Acad Sci USA* 79 (1982): 5097–5100; N. Geschwind and P. O. Behan, "Laterality, Hormones, and Immunity," in *Cerebral Dominance: The Biological Foundations*, ed. Norman Geschwind and Albert M. Galaburda (Cambridge, Mass.: Harvard University Press, 1984).

Chapter 7. The Geschwind Hypothesis

Epigraphs: D. F. Benson, "Norman Geshwind's Influence on the Study of Aphasia," in *Behavioral Neurology and the Legacy of Norman Geschwind*, ed. Steven C. Schachter and Orin Devinsky (Philadelphia: Lippincott-Raven Publishers, 1997): 71–78, 77, 71; M. Annett, "Geschwind's Legacy," *Brain Cogn* 26 (1994): 236–42, 240–41.

1. P. Bakan, G. Dibb, and P. Reed, "Handedness and Birth Stress," *Neuropsychologia* 11 (1973): 363–66.

2. Paul Bakan, "Nonright-Handedness and the Continuum of Reproductive Casualty," in *Left-Handedness: Behavioral Implications and Anamolies*, ed. Stanley Coren, Advances in Psychology, no. 67 (Oxford, England: North-Holland, 1990), 33–74, 33.

3. I. C. McManus, "Handedness and Birth Stress," *Psychol Med* 11 (1981): 485–96, 495–96.

4. N. Geschwind and P. O. Behan, "Left-Handedness: Association with Immune Disease, Migraine, and Developmental Learning Disorder," *Proc Natl Acad Sci USA* 79 (1982): 5097–5100, 5099.

5. Thomas D. Sabin, "Colleague," in Schachter and Devinsky, *Behavioral Neurology and the Legacy of Norman Geschwind*, 21–26, 23.

6. Howard Gardner, "Creative Genius," in Schachter and Devinsky, *Behavioral Neurology and the Legacy of Norman Geschwind*, 47–51, 50.

7. H. I. Kushner, "Norman Geschwind and the Use of History in the (Re)Birth of Behavioral Neurology," *J Hist Neurosci* 24 (2015): 173–92.

8. N. Geschwind, "Disconnexion Syndromes in Animals and Man. II," *Brain: A Journal of Neurology* 88 (1965): 585–644; N. Geschwind, "Disconnexion Syndromes in Animals and Man. I," *Brain: A Journal of Neurology* 88 (1965): 237–94.

9. H. I. Kushner, "Norman Geschwind and the Use of History in the (Re)Birth of Behavioral Neurology," 173–92; H. I. Kushner, "Corrigendum" (in English), *J Hist Neurosci* 24 (2015): 315.

10. For a balanced discussion of phrenology, see Anne Harrington, *Medicine, Mind, and the Double Brain: A Study in Nineteenth-Century Thought* (Princeton, N.J.: Princeton University Press, 1987), 5–8.

11. M. Catani and D. H. Ffytche, "The Rises and Falls of Disconnection Syndromes," *Brain: A Journal of Neurology* 128 (2005): 2224–39; Kushner, "Norman Geschwind and the Use of History in the (Re)Birth of Behavioral Neurology"; Geschwind, "Disconnexion Syndromes in Animals and Man. II."; Geschwind, "Disconnexion Syndromes in Animals and Man. I."

12. Catani and Ffytche, "The Rises and Falls of Disconnection Syndromes"; Benson, "Norman Geshwind's Influence on the Study of Aphasia," 74.

13. Geschwind, "Disconnexion Syndromes in Animals and Man. I," 270–75, 289–90.

14. Ibid., 275.

15. Albert M. Galaburda et al., "Right-Left Asymmetries in the Brain," *Science* 199 (1978): 852–56, 852, 855–56.

16. Geschwind and Behan, "Left-Handedness," 5099.

17. David B. Rosenfield, "Norman Geschwind (1926–1984)," in Schachter and Devinsky, *Behavioral Neurology and the Legacy of Norman Geschwind*, 101–11, 102.

18. Sabin, "Colleague," 22.

19. Ibid.

20. N. Geschwind and P. O. Behan, "Laterality, Hormones, and Immunity," in *Cerebral Dominance: The Biological Foundations*, ed. Norman Geschwind and Albert M. Galaburda (Cambridge, Mass.: Harvard University Press, 1984), 211–24, 212–13.

21. Geschwind, and Behan. "Left-Handedness," 5097–5100.

22. Ibid.

23. Ibid., 5097.

24. Ibid., 5098.

25. Ibid.

26. Geschwind, and Behan, "Laterality, Hormones, and Immunity," 211–24, 213.

27. Ibid., 214; N. Geschwind and A. M. Galaburda, *Cerebral Lateralization: Biological Mechanisms, Associations, and Pathology* (Cambridge, Mass.: MIT Press, 1987), 68–80.

28. Geschwind and Behan, "Laterality, Hormones, and Immunity," 214.

29. N. Geschwind and A. M. Galaburda, "Cerebral Lateralization: Biological Mechanisms, Associations, and Pathology: A Hypothesis and a Program for Research," *Archives of Neurology* 42, in 3 parts (1985): I: 428–59; II: 521–52; III: 634–54.

30. Ibid., III.

31. H. I. Kushner, "Evidence-Based Medicine and the Physician-Patient Dyad," *Perm J* 14 (2010): 64–69.

32. Daniel Goleman, "Left vs. Right Brain Function Tied to Hormone in the Womb," *New York Times*, September 24, 1985, 1, 16; Linda Garmon, "Of Hemispheres, Handedness, and More," *Psychology Today*, November 1985, 40–48; M. P. Bryden, I. C. McManus, and M. B. Bulman-Fleming, "Evaluating the Empirical Support for the Geschwind-Behan-Galaburda Model of Cerebral Lateralization," *Brain Cogn* 26 (November 1994): 103–67, 104.

33. Bryden, McManus, and Bulman-Fleming, "Evaluating the Empirical Support for the Geschwind-Behan-Galaburda Model of Cerebral Lateralization," 104–5.

34. Ibid., 104.

35. S. Coren and A. Searleman, "Birth Stress and Left-Handedness: The Rare Trait Marker Model," in *Left-Handedness: Behavioral Implications and Anomalies*, ed.

Stanley Coren, Advances in Psychology, no. 67 (Oxford, England: North-Holland, 1990), 3–32; Bakan, "Nonright-Handedness and the Continuum of Reproductive Casualty"; Murray Schwartz, "Left-Handedness and Prenatal Complications," in Coren, *Left-Handedness*, 75–97.

36. Garmon, "Of Hemispheres, Handedness, and More."

37. S. Coren, *The Left-Hander Syndrome: The Causes and Consequences of Left-Handedness* (New York: Free Press, 1992), 140.

38. S. Coren and D. F. Halpern, "Left-Handedness: A Marker for Decreased Survival Fitness," *Psychol Bull* 109 (1991): 90–106; D. F. Halpern and S. Coren, "Do Right-Handers Live Longer?," *Nature* 333 (1988): 213; D. F. Halpern and S. Coren, "Handedness and Life Span," *N Engl J Med* 324 (1991): 998.

39. Coren, *The Left-Hander Syndrome*, 92.

40. Ibid., 140.

41. Geschwind and Galaburda, "Cerebral Lateralization. Biological Mechanisms, Associations, and Pathology"; Geschwind and Galaburda, *Cerebral Lateralization*.

42. I. C. McManus and M. P. Bryden, "Geschwind's Theory of Cerebral Lateralization: Developing a Formal, Causal Model," *Psychol Bull* 110 (1991): 237–53, 251–52.

43. M. P. Bryden, I. C. McManus, and R. E. Steenhuis, "Handedness Is Not Related to Self-Reported Disease Incidence," *Cortex* 27 (1991): 605–11.

44. Bryden, McManus, and Bulman-Fleming, "Evaluating the Empirical Support for the Geschwind-Behan-Galaburda Model of Cerebral Lateralization," 103–67.

45. Ibid.

46. Ibid., 104.

47. Ibid., 155.

48. Ibid., 155–56. A decade later McManus had not altered his views. "On balance," he wrote in 2002, "the conclusion has to be that there is no overall support for a *simple* version of the Geschwind theory" (293–94); emphasis added. I. C. McManus, *Hypernotes to Right Hand, Left Hand: The Origins of Asymmetry in Brains, Bodies, Atoms, and Cultures*, 12:18, www.righthandlefthand.com.

49. M. Annett, "Geschwind's Legacy," *Brain and Cognition* 26 (1994): 236–42, 240.

50. Michel Habib, Florence Touze, and Albert M. Galaburda, "Intrauterine Factors in Sinistrality: A Review," in Coren, *Left-Handedness*, 92–128.

51. Coren, *Left-Handedness*, 92–128.

52. Albert M. Galaburda, telephone interview by Howard I. Kushner, July 26, 2016.

53. S. Coren, "Methodological Problems in Determining the Relationship between Handedness and Immune System Function," *Brain Cogn* 26 (1994): 168–73, 173.

54. Galaburda, interview.

Chapter 8. Genetic Models and Selective Advantage

Epigraphs: M. Annett, *Handedness and Brain Asymmetry: The Right Shift Theory* (New York: Taylor & Francis, 2002), 93; I. C. McManus, *Right Hand, Left Hand: The Origins of Asymmetry in Brains, Bodies, Atoms, and Cultures* (London: Weidenfeld & Nicolson, 2002), 229.

1. I. C. McManus, "The Inheritance of Left-Handedness," *Ciba Found Symp* 162 (1991): 267–81, 257.

2. W. M. Brandler et al., "Common Variants in Left/Right Asymmetry Genes and Pathways Are Associated with Relative Hand Skill," *PLoS Genet* 9 (2013): e1003751.

3. Annett, *Handedness and Brain Asymmetry*, 317–22.

4. J. A. Armour, A. Davison, and I. C. McManus, "Genome-Wide Association Study of Handedness Excludes Simple Genetic Models," *Heredity* (Edinburgh) 112 (March 2014): 221–25.

5. M. Annett, "Family Handedness in Three Generations Predicted by the Right Shift Theory," *Ann Hum Genet* 42 (1979): 479–91; M. Annett, "The Genetics of Handedness," *Trends in Neuroscience* 4 (1981): 256–58.

6. M. Annett, "A Classification of Hand Preference by Association Analysis," *Br J Psychol* 61 (1970): 303–21.

7. M. Annett, "Handedness in Families," *Ann Hum Genet* 37 (1973): 93–105; M. Annett, "Handedness in the Children of Two Left-Handed Parents," *Br J Psychol* 65 (1974): 129–31; M. Annett, "Hand Preference and Skill in 115 Children of Two Left-Handed Parents," *Br J Psychol* 74 (1983): 17–32.

8. Annett, *Handedness and Brain Asymmetry*, 102–6.

9. Annett, "The Genetics of Handedness"; Annett, *Handedness and Brain Asymmetry*, 93–94, 261–93.

10. M. Annett, "In Defence of the Right Shift Theory," *Perceptual and Motor Skills* 82 (1996): 115–37.

11. M. Annett et al., "The Right Shift Theory of a Genetic Balanced Polymorphism for Cerebral Dominance and Cognitive Processing. Commentaries. Author's Reply," *Cahiers de Psychologie Cognitive* 14 (1995): 427–63, 427; also see Annett, "The Genetics of Handedness"; Annett, "Hand Preference and Skill in 115 Children of Two Left-Handed Parents"; M. Annett, *Left, Right, Hand, and Brain: The Right Shift Theory* (London: Lawrence Erlbaum Associates, 1985); Annett, *Handedness and Brain Asymmetry*.

12. Annett et al., "The Right Shift Theory of a Genetic Balanced Polymorphism for Cerebral Dominance and Cognitive Processing," 427.

13. M. Annett, "Schizophrenia and Autism Considered as the Products of an Agnosic Right Shift Gene," *Cogn Neuropsychiatry* 2 (1997): 195–214, 198–99, italics in original.

14. I. C. McManus, S. Shergill, and M. P. Bryden, "Annett's Theory That Individuals Heterozygous for the Right Shift Gene Are Intellectually Advantaged: Theoretical and Empirical Problems," *Br J Psychol* 84 (1993): 517–37; Annett, "In Defence of the Right Shift Theory."

15. McManus, "The Inheritance of Left-Handedness"; McManus, *Right Hand, Left Hand*.

16. I. C. MacManus, "Precisely Wrong?: The Problems with the Jones and Martin Genetic Model of Sex Differences in Handedness and Language Lateralisation," *Cortex* 46 (2009): 700–702; M. Annett, "Left-Handedness as a Function of Sex, Maternal Versus Paternal Inheritance, and Report Bias," *Behav Genet* 29 (1999): 103–14.

17. McManus, *Right Hand, Left Hand*.

18. Ibid., 228–32.

19. Annett, *Handedness and Brain Asymmetry*, 203–16.

20. A. N. Gilbert and C. J. Wysocki, "Hand Preference and Age in the United States," *Neuropsychologia* 30 (1992): 601–8.

21. I. C. McManus et al., "Science in the Making: Right Hand, Left Hand. III: Estimating Historical Rates of Left-Handedness," *Laterality* 15 (2010): 186–208.

22. Ton G. G. Groothuis et al., "The Fighting Hypothesis in Combat: How Well Does the Fighting Hypothesis Explain Human Left-Handed Minorities?," *Ann NY Acad Sci* 1288 (2013): 100–109.

23. Charlotte Faurie and Michel Raymond, "The Fighting Hypothesis as an Evolutionary Explanation for the Handedness Polymorphism in Humans: Where Are We?," *Ann NY Acad Sci* 1288 (2013): 110–13; Groothuis et al., "The Fighting Hypothesis in Combat," 100–109.

24. P. H. Pye-Smith, "On Left-Handedness," *Guy's Hospital Reports* 16 (1871): 141–46; McManus, *Right Hand, Left Hand*, 254–55; McManus, *Hypernotes to Right Hand, Left Hand: The Origins of Asymmetry in Brains, Bodies, Atoms, and Cultures,* 10:33, www.righthandlefthand.com.

25. S. Coren, *The Left-Hander Syndrome: The Causes and Consequences of Left-Handedness* (New York: Free Press, 1992), 52–53.

26. Faurie and Raymond, "The Fighting Hypothesis as an Evolutionary Explanation for the Handedness Polymorphism in Humans."

27. Groothuis et al., "The Fighting Hypothesis in Combat," 100.

28. Michael C. Corballis and Michael J. Morgan, "On the Biological Basis of Human Laterality: I. Evidence for a Maturational Left-Right Gradient," *Behavioral and Brain Sciences* 1 (1978): 261–69, 261; N. Geschwind and A. M. Galaburda, *Cerebral Lateralization: Biological Mechanisms, Associations, and Pathology* (Cambridge, Mass.: MIT Press, 1987), 129.

29. M. C. Corballis, *The Lopsided Ape: Evolution of the Generative Mind* (New York: Oxford University Press, 1991), 202.

30. M. C. Corballis, "A House of Cards?," *Cogn Neuropsychiatry* 2 (1997): 214–16.

31. Corballis, *The Lopsided Ape*, 194.

32. Ibid.

33. Corballis, "Left Brain, Right Brain: Facts and Fantasies," *PLoS Biol* 12, no. 1 (2014): e1001767.

34. Margaret M. Clark, *Left-Handedness: Laterality Characteristics and Their Educational Implications* (London: University of London Press, 1957); G. C. Ashton, "Handedness: An Alternative Hypothesis," *Behav Genet* 12 (1982): 125–47.

Chapter 9. Uniquely Human?

Epigraphs: M. C. Corballis, *The Lopsided Ape: Evolution of the Generative Mind* (New York: Oxford University Press, 1991), 191; M. C. Corballis, "Left Brain, Right Brain: Facts and Fantasies," *PLoS Biol* 12, no. 1 (2014): e1001767.

1. Corballis, *The Lopsided Ape*.

2. M. C. Corballis, "What's Left in Language?: Beyond the Classical Model," *Ann NY Acad Sci* 1359 (2015): 14–29; Corballis, "Left Brain, Right Brain."

3. Corballis, "What's Left in Language?," 6.

4. M. C. Corballis, "Laterality and Human Evolution," *Psychol Rev* 96 (1989): 492–505, 496, 497; Corballis, *The Lopsided Ape*, 102–3.

5. Corballis, *The Lopsided Ape*, 191–92.

6. Stephen Jay Gould, "Exaptation: A Crucial Tool for an Evolutionary Psychology," *Journal of Social Issues* 47 (1991): 43–65; Stephen Jay Gould and Elisabeth S. Vrba, "Exaptation: A Missing Term in the Science of Form," *Paleobiology* 8 (1982): 4–15; Ian Tattersall, *Becoming Human: Evolution and Human Uniqueness*, 1st ed. (New York: Harcourt, Brace, & Co., 1998), 108.

7. Tattersall, *Becoming Human*, 108.

8. M. C. Corballis, "Laterality and Myth," *Am Psychol* 35 (1980): 284–95, 92.

9. Corballis, "Laterality and Human Evolution," 498.

10. Ibid., 496.

11. M. C. Corballis, "The Genetics and Evolution of Handedness," *Psychol Rev* 104 (1997): 714–27, 14.

12. M. C. Corballis, "The Evolution and Genetics of Cerebral Asymmetry," *Philos Trans R Soc Lond B Biol Sci* 364 (2009): 867–78, 875.

13. Corballis, *The Lopsided Ape*, 202.

14. Corballis, "The Genetics and Evolution of Handedness," 715–17.

15. I. C. McManus, *Right Hand, Left Hand: The Origins of Asymmetry in Brains, Bodies, Atoms, and Cultures* (London: Weidenfeld & Nicolson, 2002), 230–32.

16. M. C. Corballis, "Handedness and Cerebral Asymmetry: An Evolutionary Perspective," in *The Two Halves of the Brain: Information Processing in the Cerebral Hemispheres*, ed. K. Hugdahl and R. Westerhausen (Cambridge, Mass.: MIT Press, 2010), 65–88; Corballis, "The Evolution and Genetics of Cerebral Asymmetry"; Corballis, "Handedness and Cerebral Asymmetry"; Corballis, "Left Brain, Right Brain."

17. William C. McGrew, Wulf Schiefenhövel, and Linda F. Marchant, "Introduction to the Evolution of Human Handedness," *Ann NY Acad Sci* 1288 (2013): v–vi.

18. Ibid.

19. Ibid.

20. William D. Hopkins, "Neuroanatomical Asymmetries and Handedness in Chimpanzees (Pan Troglodytes): A Case for Continuity in the Evolution of Hemispheric Specialization," *Ann NY Acad Sci* 1288 (2013): 17–35, 17.

21. Jeroen B. Smaers et al., "Laterality and the Evolution of the Prefronto-Cerebellar System in Anthropoids," *Ann NY Acad Sci* 1288 (2013): 59–69, 59.

22. Jay T. Stock et al., "Skeletal Evidence for Variable Patterns of Handedness in Chimpanzees, Human Hunter-Gatherers, and Recent British Populations," *Ann NY Acad Sci* 1288 (2013): 86–99, 86; also see Janina Tutkuviene and Wulf Schiefenhövel, "Laterality of Handgrip Strength: Age- and Physical Training–Related Changes in Lithuanian Schoolchildren and Conscripts," *Ann NY Acad Sci* 1288 (2013): 124–34.

23. Linda F. Marchant and William C. McGrew, "Handedness Is More Than Laterality: Lessons from Chimpanzees," *Ann NY Acad Sci* 1288 (2013): 1–8, 1.

24. W. Tecumseh Fitch and Stephanie N. Braccini, "Primate Laterality and the Biology and Evolution of Human Handedness: A Review and Synthesis," *Ann NY Acad Sci* 1288 (2013): 70–85, 70; also see Catherine Hobaiter and Richard W. Byrne, "Laterality in the Gestural Communication of Wild Chimpanzees," *Ann NY Acad Sci* 1288 (2013): 9–16.

25. W. D. Hopkins et al., "Behavioral and Brain Asymmetries in Primates: A Preliminary Evaluation of Two Evolutionary Hypotheses," *Ann NY Acad Sci* 359 (2015):

65–83; also see A. Meguerditchian, J. Vauclair, and W. D. Hopkins, "On the Origins of Human Handedness and Language: A Comparative Review of Hand Preferences for Bimanual Coordinated Actions and Gestural Communication in Nonhuman Primates," *Dev Psychobiol* 55 (2013): 637–50.

26. Clare Porac, *Laterality: Exploring the Enigma of Left-Handedness* (London: Academic Press, 2016), 210.

27. Jacqueline Fagard, "The Nature and Nurture of Human Infant Hand Preference," *Ann NY Acad Sci* 1288 (2013): 114–23, 114.

28. Thomas H. Priddle and Timothy J. Crow, "The Protocadherin 11x/Y (Pcdh11x/Y) Gene Pair as Determinant of Cerebral Asymmetry in Modern Homo Sapiens," *Ann NY Acad Sci* 1288 (2013): 36–47.

29. I. C. McManus, Angus Davison, and John A. L. Armour, "Multilocus Genetic Models of Handedness Closely Resemble Single-Locus Models in Explaining Family Data and Are Compatible with Genome-Wide Association Studies," *Ann NY Acad Sci* 1288 (2013): 48–58, 48.

30. Charlotte Faurie et al., "Variation in the Frequency of Left-Handedness in Traditional Societies," *Current Anthropology* 46 (2005): 142–47.

31. Charlotte Faurie and Michel Raymond, "The Fighting Hypothesis as an Evolutionary Explanation for the Handedness Polymorphism in Humans: Where Are We?," *Ann NY Acad Sci* 1288 (2013): 110–13, 110.

32. C. Faurie et al., "Left-Handedness and Male-Male Competition: Insights from Fighting and Hormonal Data," *Evol Psychol* 9 (2011): 354–70, 354.

33. Ton G. G. Groothuis et al., "The Fighting Hypothesis in Combat: How Well Does the Fighting Hypothesis Explain Human Left-Handed Minorities?," *Ann NY Acad Sci* 1288 (2013): 100–109, 100, 107.

34. W. M. Brandler et al., "Common Variants in Left/Right Asymmetry Genes and Pathways Are Associated with Relative Hand Skill," *PLoS Genet* 9 (2013): e1003751; W. M. Brandler and S. Paracchini, "The Genetic Relationship between Handedness and Neurodevelopmental Disorders," *Trends Mol Med* 20 (2014): 83–90.

35. W. D. Hopkins et al., "Hand Preferences for Coordinated Bimanual Actions in 777 Great Apes: Implications for the Evolution of Handedness in Hominins," *J Hum Evol* 60 (2011): 605–11; W. D. Hopkins et al., "Genetic Basis in Motor Skill and Hand Preference for Tool Use in Chimpanzees (Pan Troglodytes)," *Proc Biol Sci* 282, no. 1800 (2015): 20141223; Hopkins et al., "Behavioral and Brain Asymmetries in Primates."

36. Lindsay M. Oberman, Jaime A. Pineda, and Vilayanur S. Ramachandran, "The Human Mirror Neuron System: A Link between Action Observation and Social Skills," *Social Cognitive and Affective Neuroscience* 1 (2007): 62–66.

37. Andrey Giljov et al., "Parallel Emergence of True Handedness in the Evolution of Marsupials and Placentals," *Current Biology* 25 (2015): 1878–84.

38. Corballis, "The Evolution and Genetics of Cerebral Asymmetry," 867; also see Corballis, "Left Brain, Right Brain."

39. Corballis, "Left Brain, Right Brain."

40. Ibid., 1.

41. Giljov et al., "Parallel Emergence of True Handedness in the Evolution of Marsupials and Placentals."

42. Ibid., 5; Michael Corballis to Howard Kushner, August 6, 2015, private communication.

43. Corballis, "Laterality and Myth," 292. In a fascinating book, Ian McGilchrist has come to a similar but opposite conclusion, arguing that the left hemisphere hijacked right-hemispheric dominance, resulting in an emphasis on the individual rather than the collective. I. McGilchrist, *The Master and His Emissary: The Divided Brain and the Making of the Western World* (Yale University Press, 2009).

44. Corballis, "What's Left in Language?," 1.

45. Ibid.

46. G. Rizzolatti, "Multiple Body Representations in the Motor Cortex of Primates," *Acta Biomed Ateneo Parmense* 63 (1992): 27–29; Oberman, Pineda, and Ramachandran, "The Human Mirror Neuron System," 62–66; Katja Guenther, "Imperfect Reflections: Norms, Pathology, and Difference in Mirror Neuron Research," in *Plasticity and Pathology: On the Formation of the Neural Subject*, ed. David Bates and Nima Bassiri (New York: Fordham University Press, 2016): 268–308.

47. Lindsay M. Oberman, Vilayanur S. Ramachandran, and Jaime A. Pineda, "Modulation of Mu Suppression in Children with Autism Spectrum Disorders in Response to Familiar or Unfamiliar Stimuli: The Mirror Neuron Hypothesis," *Neuropsychologia* 46 (2008): 1558–65; Michael A. Arbib, *How the Brain Got Language: The Mirror System Hypothesis*, Oxford Studies in the Evolution of Language (Oxford, UK: Oxford University Press, 2012).

48. Corballis, "What's Left in Language?," 5–6.

49. I. S. Haberling, P. M. Corballis, and M. C. Corballis, "Language, Gesture, and Handedness: Evidence for Independent Lateralized Networks," *Cortex* 82 (2016): 72–85, 72.

50. F. B. M. de Waal, *The Age of Empathy: Nature's Lessons for a Kinder Society* (New York: Harmony Books, 2009).

51. G. Hickok, *The Myth of Mirror Neurons: The Real Neuroscience of Communication and Cognition*, 1st ed. (New York: Norton, 2014); Allan Young, "The Social Brain and the Myth of Empathy," *Science in Context* 25 (2012): 410–24; G. Hickok et al., "The Role of Broca's Area in Speech Perception: Evidence from Aphasia Revisited," *Brain Lang* 119 (2011): 214–20; G. Hickok, "The Role of Mirror Neurons in Speech and Language Processing," *Brain Lang* 112 (2010): 1–2, 1; K. Emmorey, "The Neurobiology of Sign Language and the Mirror System Hypothesis," *Lang Cogn* 5 (2013): 205–10.

52. Guenther, "Imperfect Reflections," 269.

53. Emmorey, "The Neurobiology of Sign Language and the Mirror System Hypothesis," 206.

54. Hickok et al., "The Role of Broca's Area in Speech Perception."

55. Corballis, "What's Left in Language?" His revised hypothesis, writes Corballis, "provides a more Darwinian perspective on language and its lateralization" (1). The utility of this claim, of course, rests upon how one defines "Darwinian."

56. Corballis, "Left Brain, Right Brain," 1.

57. Ibid., 4.

58. N. Geschwind and A. M. Galaburda, *Cerebral Lateralization: Biological Mechanisms, Associations, and Pathology* (Cambridge, Mass.: MIT Press, 1987).

59. Murray Schwartz, "Left-Handedness and Prenatal Complications," in *Left-Handedness: Behavioral Implications and Anomalies,* ed. Stanley Coren, Advances in Psychology, no. 67 (Oxford, England: North-Holland, 1990), 75–97, 92.

CHAPTER 10. A GAY HAND?

Epigraphs: N. Geschwind and A. M. Galaburda, *Cerebral Lateralization: Biological Mechanisms, Associations, and Pathology* (Cambridge, Mass.: MIT Press, 1987), 175; Robert Williams (pseudonym), email to Howard Kushner, July 18, 2016.

1. Nicole Bode, "Sexuality at Hand: Left-Handers Are More Likely to Be Gay. Is There a Genetic Link?" *Psychology Today,* November 1, 2000.

2. M. L. Lalumière, R. Blanchard, and K. J. Zucker, "Sexual Orientation and Handedness in Men and Women: A Meta-Analysis," *Psychol Bull* 126 (2000): 575–92, 575. Emphasis added.

3. Cesare Lombroso, *Criminal Man,* ed. Mary Gibson and Nicole Hahn Rafter, trans. Mary Gibson and Nicole Hahn Rafter (Durham: Duke University Press, 2006), 273–76.

4. Cesare Lombroso, "Left-Handedness and Left-Sidedness," *North American Review* 177 (1903): 440–44.

5. S. Coren and A. Searleman, "Birth Stress and Left-Handedness: The Rare Trait Marker Model," in *Left-Handedness: Behavioral Implications and Anomalies,* ed. Stanley Coren, Advances in Psychology, no. 67 (Oxford, England: North-Holland, 1990), 3–32, 4–5; I. C. McManus, *Right Hand, Left Hand: The Origins of Asymmetry in Brains, Bodies, Atoms, and Cultures* (London: Weidenfeld & Nicolson, 2002), 35, 152–53.

6. Geschwind and Galaburda, *Cerebral Lateralization,* 175.

7. Williams, email to Howard Kushner, July 18, 2016.

8. L. D. Rosenstein and E. D. Bigler, "No Relationship between Handedness and Sexual Preference," *Psychol Rep* 60 (1987): 704–6.

9. B. A. Gladue and J. M. Bailey, "Spatial Ability, Handedness, and Human Sexual Orientation," *Psychoneuroendocrinology* 20 (1995): 487–97.

10. A. F. Bogaert and R. Blanchard, "Handedness in Homosexual and Heterosexual Men in the Kinsey Interview Data," *Arch Sex Behav* 25 (1996): 373–78. The investigators did report that both groups had predicted rates of *non-right-handedness* (11%–12%), a category whose robustness was discussed in chapter 3 and to which we will return below.

11. S. LeVay and D. H. Hamer, "Evidence for a Biological Influence in Male Homosexuality," *Sci Am* 270 (1994): 44–49; S. LeVay, "A Difference in Hypothalamic Structure between Heterosexual and Homosexual Men," *Science* 253 (1991): 1034–37.

12. S. Coren, *The Left-Hander Syndrome: The Causes and Consequences of Left-Handedness* (New York: Free Press, 1992), 201.

13. Clare Porac, *Laterality: Exploring the Enigma of Left-Handedness* (London: Academic Press, 2016), 80.

14. Zucker became editor-in-chief of the *Archives of Sexual Behavior* (the official publication of the International Academy of Sex Research) in 2001; http://link.springer.com/journal/10508.

15. Lalumière, Blanchard, and Zucker, "Sexual Orientation and Handedness in Men and Women," 574.

16. Simon LeVay was skeptical, noting that "the data obscures the fact that most homosexuals are right-handed, and most left-handed people are heterosexual." Bode, "Sexuality at Hand."

17. Lalumière, Blanchard, and Zucker, "Sexual Orientation and Handedness in Men and Women," 575. In a follow-up study, Zucker concluded that "left-handedness appears to be a behavioral marker of an underlying neurobiological process associated with gender identity disorder in boys." K. J. Zucker et al., "Handedness in Boys with Gender Identity Disorder," *J Child Psychol Psychiatry* 42 (2001): 30–35.

18. Lalumière, Blanchard, and Zucker, "Sexual Orientation and Handedness in Men and Women," 578. "Women with a gender identity disorder, and married men with a gender identity disorder, daughters of mothers exposed to diethylstilbestrol (DES) during pregnancy, and female patients with congenital adrenal hyperplasia (CAH) were not included in the meta-analysis but were examined separately" (578).

19. Lalumière, Blanchard, and Zucker, "Sexual Orientation and Handedness in Men and Women," 775.

20. I. C. McManus, *Hypernotes to Right Hand, Left Hand: The Origins of Asymmetry in Brains, Bodies, Atoms, and Cultures*, 7:10, www.righthandlefthand.com.

21. M. Kishida and Q. Rahman, "Fraternal Birth Order and Extreme Right-Handedness as Predictors of Sexual Orientation and Gender Nonconformity in Men," *Arch Sex Behav* 44, no. 5 (July 2015): 1493–501, 1493. This was an inadvertent finding. The researchers were testing the hypothesis that there was a relationship between the number of older brothers and an increased risk of homosexuality in younger siblings. Their data failed to support this hypothesis.

22. M. A. Yule, L. A. Brotto, and B. B. Gorzalka, "Biological Markers of Asexuality: Handedness, Birth Order, and Finger Length Ratios in Self-Identified Asexual Men and Women," *Arch Sex Behav* 43 (2014): 299–310.

23. McManus, *Hypernotes to Right Hand, Left Hand*, 7:10.

24. J. Preslar et al., "Autism, Lateralisation, and Handedness: A Review of the Literature and Meta-Analysis," *Laterality* 19 (2014): 64–95.

25. Rosenstein and Bigler, "No Relationship between Handedness and Sexual Preference."

26. W. F. Daniel and R. A. Yeo, "Handedness and Sexual Preference: A Re-Analysis of Data Presented by Rosenstein and Bigler," *Percept Mot Skills* 76 (1993): 544–46.

27. L. M. Lansky, H. Feinstein, and J. M. Peterson, "Demography of Handedness in Two Samples of Randomly Selected Adults (N = 2083)," *Neuropsychologia* 26 (1988): 465–77.

28. Lalumière, Blanchard, and Zucker, "Sexual Orientation and Handedness in Men and Women," 580–82, 580.

29. S. D. Stellman, E. L. Wynder, D. J. DeRose, and J. E. Muscat, "The Epidemiology of Left-Handedness in a Hospital Population," *Ann Epidemiol* 7, no. 3 (April 1997): 167–71. Also see Lalumière, Blanchard, and Zucker, "Sexual Orientation and Handedness in Men and Women," 585.

30. S. E. Marchant-Haycox, I. C. McManus, and G. D. Wilson, "Left-Handedness, Homosexuality, HIV Infection and Aids," *Cortex* 27 (1991): 49–56.

CHAPTER 11. DISABILITY, ABILITY, AND THE LEFT HAND

Epigraphs: I. C. McManus, *Right Hand, Left Hand: The Origins of Asymmetry in Brains, Bodies, Atoms, and Cultures* (London: Weidenfeld & Nicolson, 2002), 230–31; D. V. Bishop, *Handedness and Developmental Disorder* (Philadelphia: Blackwell Scientific, 1990), 163.

1. Cesare Lombroso, "Left-Handedness and Left-Sidedness," *North American Review* 177 (1903): 440–44; Robert Hertz, "La Prééminence de la Main Doite: Étude Sur la Polarité Religieuse," *Revue Philosophique* LXVIII (1909): 553–80.

2. G. LeBon, *L'homme et les sociétés, Leurs origines et leur histoire* (1881) (Paris: Jean-Michel Place, 1988); Anne Harrington, *Medicine, Mind, and the Double Brain: A Study in Nineteenth-Century Thought* (Princeton, N.J.: Princeton University Press, 1987), 88–89, 95–99; Howard I. Kushner, "Suicide, Gender, and the Fear of Modernity in Nineteenth-Century Medical and Social Thought," *Journal of Social History* 26 (1993): 461–90, 471–73.

3. John Jackson, *Ambidexterity: or, Two-Handedness and Two-Brainedness: An Argument for Natural Development and Rational Education* (London: Kegan Paul, Tench, Trubner & Co., 1905; reprinted 2009 by Bibliolife), xi; Daniel Wilson, *The Right Hand: Left Handedness* (New York: McMillan, 1891).

4. M. Annett, *Handedness and Brain Asymmetry: The Right Shift Theory* (New York: Taylor & Francis, 2002).

5. Annukka K. Lindell, "On the Interrelation between Reduced Lateralization, Schizotypy, and Creativity," *Frontiers in Psychology* 5 (2014): 813, 1–4; A. K. Lindell, "Lateral Thinkers Are Not So Laterally Minded: Hemispheric Asymmetry, Interaction, and Creativity," *Laterality* 16 (2011): 479–98.

6. T. J. Crow, "A Continuum of Psychosis, One Human Gene, and Not Much Else—The Case for Homogeneity," *Schizophr Res* 17 (1995): 135–45; T. J. Crow, "Aetiology of Schizophrenia: An Evolutionary Theory," *Int Clin Psychopharmacol* 10 Suppl 3 (1995): 49–56.

7. M. Annett, "Schizophrenia and Autism Considered as the Products of an Agnosic Right Shift Gene," *Cogn Neuropsychiatry* 2 (1997): 195–214; Annett, *Handedness and Brain Asymmetry*, 198.

8. Annett, "Schizophrenia and Autism Considered as the Products of an Agnosic Right Shift Gene," 198–99.

9. Thus, the *RS*+ codes for left-brain asymmetry; the *RS*– for symmetry between the hemispheres; while the *RS*+a for either asymmetry or symmetry by chance.

10. D. Leonhard and P. Brugger, "Creative, Paranormal, and Delusional Thought: A Consequence of Right Hemisphere Semantic Activation?," *Neuropsychiatry Neuropsychol Behav Neurol* 11 (1998): 177–83, 177.

11. Lindell, "On the Interrelation between Reduced Lateralization, Schizotypy, and Creativity," 813, 3.

12. M. C. Corballis, "A House of Cards?," *Cogn Neuropsychiatry* 2 (1997): 214–16, 215–16.

13. I. C. McManus, "Autism and Schizophrenia Are Not Due to a Singlegene Locus," *Cogn Neuropsychiatry* 2 (1997): 226–31.

14. Ibid.

15. M. Annett, "Schizophrenia and Autism Considered as the Products of an Agnosic Right Shift Gene," 207–8, italics in original.

16. I. Sommer et al., "Handedness, Language Lateralisation and Anatomical Asymmetry in Schizophrenia: Meta-Analysis," *Br J Psychiatry* 178 (2001): 344–51.

17. M. Somers et al., "Hand-Preference and Population Schizotypy: A Meta-Analysis," *Schizophr Res* 108 (2009): 25–32.

18. M. Dragovic and G. Hammond, "Handedness in Schizophrenia: A Quantitative Review of Evidence," *Acta Psychiatr Scand* 111 (2005): 410–19.

19. Going further, they asserted that atypical handedness posed a risk for "normal" subjects who also scored high on the left- and mixed-handedness index. A. Preti, C. Sardu, and A. Piga, "Mixed-Handedness Is Associated with the Reporting of Psychotic-Like Beliefs in a Non-Clinical Italian Sample," *Schizophr Res* 92 (2007): 15–23.

20. Anouk van der Hoorn et al., "Non-Right-Handedness and Mental Health Problems among Adolescents from the General Population: The Trails Study," *Laterality* 15 (2009): 304–16.

21. D. W. Johnston et al., "Nature's Experiment?: Handedness and Early Childhood Development," *Demography* 46 (2009): 281–301.

22. M. Alary et al., "Functional Hemispheric Lateralization for Language in Patients with Schizophrenia," *Schizophr Res* 149 (2013): 42–47.

23. K. A. Dorph-Petersen et al., "The Influence of Chronic Exposure to Antipsychotic Medications on Brain Size before and after Tissue Fixation: A Comparison of Haloperidol and Olanzapine in Macaque Monkeys," *Neuropsychopharmacology* 30 (2005): 1649–61.

24. S. Ocklenburg et al., "Laterality and Mental Disorders in the Postgenomic Age: A Closer Look at Schizophrenia and Language Lateralization," *Neurosci Biobehav Rev* 59 (2015): 100–110.

25. H. Goez and N. Zelnik, "Handedness in Patients with Developmental Coordination Disorder," *J Child Neurol* 23 (2008): 151–54.

26. A. L. Rysstad and A. V. Pedersen, "Brief Report: Non-Right-Handedness within the Autism Spectrum Disorder," *J Autism Dev Disord* 46 (2016): 1110–17. Most recently, see T. A. Knaus, J. Kamps, and A. L. Foundas, "Handedness in Children with Autism Spectrum Disorder," *Percept Mot Skills* 122 (2016): 542–59.

27. J. Preslar et al., "Autism, Lateralisation, and Handedness: A Review of the Literature and Meta-Analysis," *Laterality* 19 (2014): 64–95, 64.

28. Lindell, "Lateral Thinkers Are Not So Laterally Minded: Hemispheric Asymmetry, Interaction, and Creativity," 479–98.

29. McManus, *Right Hand, Left Hand*, 226–27.

30. Bishop, *Handedness and Developmental Disorder*, 163.

31. D. V. Bishop, "Cerebral Asymmetry and Language Development: Cause, Correlate, or Consequence?," *Science* 340 (2013): 1230531, 2, 5–6.

CONCLUSION

Epigraph: Clare Porac, *Laterality: Exploring the Enigma of Left-Handedness* (London: Academic Press, 2016), 211.

1. Albert M. Galaburda, interview by Howard I. Kushner, July 26, 2016; Galaburda, email to Howard Kushner, December 13, 2016.

2. Kenneth Heilman, email to Howard Kushner, December 13, 2016.

3. Galaburda, interview by Howard I. Kushner, July 26, 2016; Galaburda, email to Howard Kushner, December 13, 2016.

4. W. M. Brandler et al., "Common Variants in Left/Right Asymmetry Genes and Pathways Are Associated with Relative Hand Skill," *PLoS Genet* 9 (2013): e1003751.

5. Y. K. Chan and P. S. Loh, "Handedness in Man: The Energy Availability Hypothesis," *Med Hypotheses* 94 (2016): 108–11.

6. Charlotte Faurie and Michel Raymond, "The Fighting Hypothesis as an Evolutionary Explanation for the Handedness Polymorphism in Humans: Where Are We?," *Ann NY Acad Sci* 1288 (2013): 110–13.

7. S. Coren, *The Left-Hander Syndrome: The Causes and Consequences of Left-Handedness* (New York: Free Press, 1992).

8. D. V. Bishop, "Cerebral Asymmetry and Language Development: Cause, Correlate, or Consequence?," *Science* 340, no. 6138 (June 14, 2013): 1230531.

9. Galaburda, interview and email.

10. McManus questions the claim that da Vinci was left-handed. I. C. McManus, *Right Hand, Left Hand: The Origins of Asymmetry in Brains, Bodies, Atoms, and Cultures* (London: Weidenfeld & Nicolson, 2002). 318–20; "Top 10 Lefties," *Time*, Aug. 13, 2014, http://time.com/3107557/top-10-lefties/.

11. WinCalendar, http://www.wincalendar.com/International-Left-Handers-Day; Michael Cavna, "On Left-Handers Day, a New Bill of Rights—or Lefts—for Aggrieved Southpaws," *Washington Post*, June 13, 2016.

12. The claims that musicians, visual artists, and creative writers are more likely to be lefties, Heilman concludes, are not supported by persuasive evidence. Kenneth M. Heilman, *Creativity and the Brain* (New York: Psychology Press, 2005), 73–88.

13. Les Gauchers, "L'école Des Années 50 et Les Gauchers," *Gauchers, Journale Internationale des Gauchers*, no. 13 (June 2014).

14. John L. Dawson, "Temne-Arunta Hand-Eye Dominance and Cognitive Style," *International Journal of Psychology* 7 (1972): 219–33.

15. Michael Hackh, August 1, 2016, http://linkshaenderforum.org/forum/index.php.

Index